VIRTUAL REALITY

AUGMENTED REALITY
MIXED REALITY

HUMAN-COMPUTER INTERACTION

FEATURED WITH
- D'FUSION STUDIO
- OCULUS RIFT
- GOOGLE GLASS
- HOLOLENS

 AJIT SINGH

PREFACE

No matter how many different size and shape the computer has, the basic components of computers are still the same. If we use the user perspective to look for the development of computer history, we can surprisingly find that it is the input output device that leads the development of the industry development, in one word, human-computer interaction changes the development of computer history.

Human Computer Interaction has been gone through three stages, the first stage relies on the input command to interact with the computer and the main input device is the keyboard. The second stage, which occurs presently, is the graphical user interface and the main input device is the mouse. The third stage is the touch screen which is used in every smartphone and tablet and the main interaction at this stage is finger. There is one clear point in such history, the basic evolution direction is to let people have more freedom, bring a more natural way to gain information, people want to fully use the human five senses to feel the world, thus virtual reality and augmented reality has become a hot topic.

To augment reality is to alter the view of the physical world through use of computer-generated sensory and image processing. It is the combination of reality and virtual reality, and acts as a technological extension to our own vision. The uses of augmented reality range from displaying critical information about a patient during surgery, to showing which the highest rated restaurant in a food court is. Information being overlaid on top of reality in real-time could drastically improve the efficiency and effectiveness of almost any every-day activity. This technology is still at its breaking through point and has much progress to make. This book will only skim the surface of augmented reality technology, but still make an effective use of the technology on a small rover.

This book will introduce the history and development of human–computer interaction and Virtual Reality / Augmented Reality /Mixed Reality. It will analyze three most representive technologies, Oculus Rift, Google Glass and HoloLens. Based on these, it will discuss the advantages and disadvantages of those technologies. This book discusses how VR, AR and MR work, and provides reasoning why virtual reality and augmented reality will be the next stage of human-computer interaction and how much possibility there is that VR, AR and MR technologies will be the next stage of HCI.

The book contains the following chapters:
Chapter 1 introduces the background of the book.
Chapter 2 describes the development of Human–Computer Interaction.
Chapter 3 introduces Virtual Reality and analyses some of the VR technologies, like Oculus Rift.
Chapter 4 introduces Augmented Reality and analyses Google Glass and HoloLens.
Chapter 5 is the Mixed Reality
Chapter 6 Implementation Augmented Reality In Learning
Chapter 7 DIY Creating Augmented Reality With D'FUSION

CONTENTS

1 INTRODUCTION

With the rapid development of modern technology, technology is everywhere. The computer is the most representative product, as within a few decades, now there are many different types of computers, for example, the huge server located in the room, the personal computer on the table, the laptop on the knees, the smartphone and tablet in our hands, even the wearable device on our wrist or on our head. The relationship between the computer and human has gone through a fundamental change. Human–Computer Interaction (HCI) is a field of study that helps people control the machines more easily, so that computers can be used by a range of users, from the minority specialist to the majority of people around the world. ACM defines HCI as "A discipline concerned with the design, evaluation and implementation of interactive computing systems for human use and with the study of major phenomena surrounding them." (Hewett et al., 2009).

At first, we could only type the command to let the computer work, then after the development of the graphical user interface, with the birth of the mouse, we started to click the icon to work. In the recent years, the technology of touch screen has freed our hand and we can touch the screen to use our device. People still want to find more ways of interaction with the computer. As one of the main types of Human–Computer Interaction Virtual Reality has become a hot topic in the recent years. Oculus Rift, Google Glass, and HoloLens represent the most advanced technology in the field of Virtual Reality / Augmented Reality (AR). The communication between human beings and the real world is quite normal, the virtual world is over isolated, so is there a communication platform between reality and virtuality which helps people find more friends?

People just repeat tedious everyday life activities and feel bored. Is there a way that can make the real world full of variety? AR technology can solve the above problems, it is a technology that can improve and fulfill new demands and it will be the next big thing. This book will introduce the history and development of human–computer interaction and Virtual Reality / Augmented Reality. It will analyze three most representive technologies, Oculus Rift, Google Glass and HoloLens. Based on these, it will discuss the advantages and disadvantages of those technologies, and will conclude on how much possibility there is that VR and AR technology will be the next stage of HCI.

1.1.　History

Nowadays computer graphics is used in many domains of our life. At the end of the 20th century it is difficult to imagine an architect, engineer, or interior designer working without a graphics workstation. In the last years the stormy development of microprocessor technology brings faster and faster computers to the market. These machines are equipped with better and faster graphics boards and their prices fall down rapidly. It becomes possible even for an average user, to move into the world of computer graphics. This fascination with a new (ir)reality often starts with computer games and lasts forever. It allows to see the surrounding world in other dimension and to experience things that are not accessible in real life or even not yet created. Moreover, the world of three-dimensional graphics has neither borders nor constraints and can be created and manipulated by ourselves as we wish – we can enhance it by a fourth dimension: the dimension of our imagination.

But not enough: people always want more. They want to step into this world and interact with it – instead of just watching a picture on the monitor. This technology which becomes overwhelmingly popular and fashionable in current decade is called *Virtual Reality* (VR). The very first idea of it was presented by Ivan Sutherland in 1965: "make that (virtual) world in the window look real, sound real, feel real, and respond realistically to the viewer's actions" [Suth65]. It has been a long time since then, a lot of research has been done and status quo: "the Sutherland's challenge of the

Promised Land has not been reached yet but we are at least in sight of it".

Let us have a short glimpse at the last three decades of research in virtual reality and its highlights:

Sensorama – in years 1960-1962 Morton Heilig created a multi-sensory simulator. A prerecorded film in color and stereo, was augmented by binaural sound, scent, wind and vibration experiences. This was the first approach to create a virtual reality system and it had all the features of such an environment, but it was not interactive.

The Ultimate Display – in 1965 Ivan Sutherland proposed the ultimate solution of virtual reality: an artificial world construction concept that included interactive graphics, force-feedback, sound, smell and taste.

"The Sword of Damocles" – the first virtual reality system realized in hardware, not in concept. Ivan Sutherland constructs a device considered as the first *Head Mounted Display* (HMD), with appropriate head tracking. It supported a stereo view that was updated correctly according to the user's head position and orientation.

GROPE – the first prototype of a force-feedback system realized at the University of North Carolina (UNC) in 1971.

VIDEOPLACE – Artificial Reality created in 1975 by Myron Krueger – "a conceptual environment, with no existence". In this system the silhouettes of the users grabbed by the cameras were projected on a large screen. The participants were able to interact one with the other thanks to the image processing techniques that determined their positions in 2D screen's space.

VCASS – Thomas Furness at the US Air Force's Armstrong Medical Research Laboratories developed in 1982 the Visually Coupled Airborne Systems Simulator – an advanced flight simulator. The fighter pilot wore a HMD that augmented the out-the-window view by the graphics describing targeting or optimal flight path information.

VIVED – VIrtual Visual Environment Display – constructed at the NASA Ames in 1984 with off-the-shelf technology a stereoscopic monochrome HMD.

VPL – the VPL company manufactures the popular DataGlove (1985) and the Eyephone HMD (1988) – the first commercially available VR devices.

BOOM – commercialized in 1989 by the Fake Space Labs. BOOM is a small box containing two CRT monitors that can be viewed through the eye holes. The user can grab the box, keep it by the eyes and move through the virtual world, as the mechanical arm measures the position and orientation of the box.

UNC Walkthrough project – in the second half of 1980s at the University of North Carolina an architectural walkthrough application was developed. Several VR devices were constructed to improve the quality of this system like: HMDs, optical trackers and the Pixel-Plane graphics engine.

Virtual Wind Tunnel – developed in early 1990s at the NASA Ames application that allowed the observation and investigation of flow-fields with the help of BOOM and DataGlove.

6

CAVE – presented in 1992 CAVE (CAVE Automatic Virtual Environment) is a virtual reality and scientific visualization system. Instead of using a HMD it projects stereoscopic images on the walls of room (user must wear LCD shutter glasses). This approach assures superior quality and resolution of viewed images, and wider field of view in comparison to HMD based systems.

Augmented Reality (AR) – a technology that "presents a virtual world that enriches, rather than replaces the real world" [Brys92c]. This is achieved by means of see-through HMD that superimposes virtual three-dimensional objects on real ones. This technology was previously

1.2.   What is VR? What is VR not?

At the beginning of 1990s the development in the field of virtual reality became much more stormy and the term Virtual Reality itself became extremely popular. We can hear about Virtual Reality nearly in all sort of media, people use this term very often and they misuse it in many cases too. The reason is that this new, promising and fascinating technology captures greater interest of people than e.g., computer graphics. The consequence of this state is that nowadays the border between 3D computer graphics and Virtual Reality becomes fuzzy. Therefore in the following sections some definitions of Virtual Reality and its basic principles are presented.

1.2.1. Some basic definitions and terminology

Virtual Reality (VR) and *Virtual Environments* (VE) are used in computer community interchangeably. These terms are the most popular and most often used, but there are many other. Just to mention a few most important ones: *Synthetic Experience*, *Virtual Worlds, Artificial Worlds* or *Artificial Reality*. All these names mean the same:

"Real-time interactive graphics with three-dimensional models, combined with a display technology that gives the user the immersion in the model world and direct manipulation."

"The illusion of participation in a synthetic environment rather than external observation of such an environment. VR relies on a three-dimensional, stereoscopic head-tracker displays, hand/body tracking and binaural sound. VR is an immersive, multi-sensory experience."

"Computer simulations that use 3D graphics and devices such as the DataGlove to allow the user to interact with the simulation."

"Virtual reality refers to immersive, interactive, multi-sensory, viewer-centered, three-dimensional computer generated environments and the combination of technologies required to build these environments."

"Virtual reality lets you navigate and view a world of three dimensions in real time, with six degrees of freedom. (…) In essence, virtual reality is clone of physical reality."

Although there are some differences between these definitions, they are essentially equivalent. They all mean that VR is an interactive and immersive (with the feeling of presence) experience in a simulated (autonomous) world – and this measure we will use to determine the level of advance of VR systems.

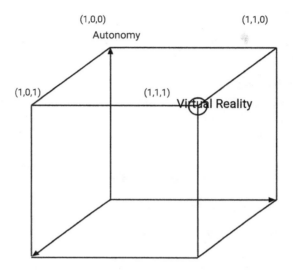

Figure 1.2.1.1. Autonomy, interaction, presence in VR – Zeltzer's cube.

Many people, mainly the researchers use the term Virtual Environments instead of Virtual Reality "because of the hype and the associated unrealistic expectations". Moreover, there are two important terms that must be mentioned when talking about VR: *Telepresence* and *Cyberspace*. They are both tightly coupled with VR, but have a slightly different context:

Telepresence – is a specific kind of virtual reality that simulates a real but remote (in terms of distance or scale) environment. Another more precise definition says that telepresence occurs when "at the work site, the manipulators have the dexterity to allow the operator to perform normal human functions; at the control station, the operator receives sufficient quantity and quality of sensory feedback to provide a feeling of actual presence at the worksite".

Cyberspace – was invented and defined by William Gibson as "a consensual hallucination experienced daily by billions of legitimate operators (...) a graphics representation of data abstracted from the banks of every computer in human system" [Gibs83]. Today the term Cyberspace is rather associated with entertainment systems and World Wide Web (Internet).

1.2.2. Levels of immersion in VR systems

In a virtual environment system a computer generates sensory impressions that are delivered to the human senses. The type and the quality of these impressions determine the level of immersion and the feeling of presence in VR. Ideally the high-resolution, high-quality and consistent over all the displays, information should be presented to all of the user's senses. Moreover, the environment itself should react realistically to the user's actions. The practice, however, is very different from this ideal case. Many applications stimulate only one or a few of the senses, very often with low-quality and unsynchronized information. We can group the VR systems accordingly to the level of immersion they offer to the user:

Desktop VR – sometimes called Window on World (WoW) systems. This is the simplest type of virtual reality applications. It uses a conventional monitor to display the image (generally monoscopic) of the world. No other sensory output is supported.

Fish Tank VR – improved version of Desktop VR. These systems support head tracking and therefore improve the feeling of "of being there" thanks to the motion parallax effect. They still use a conventional monitor (very often with LCD shutter glasses for stereoscopic viewing) but generally do not support sensory output.

Immersive systems – the ultimate version of VR systems. They let the user totally immerse in computer generated world with the help of HMD that supports a stereoscopic view of the scene accordingly to the user's position and orientation. These systems may be enhanced by audio, haptic and sensory.

1.3. Applications of VR

1.3.1. Motivation to use VR

Undoubtedly VR has attracted a lot of interest of people in last few years. Being a new paradigm of user interface it offers great benefits in many application areas. It provides an easy, powerful, intuitive way of human-computer interaction. The user can watch and manipulate the simulated environment in the same way we act in the real world, without any need to learn how the complicated (and often clumsy) user interface works. Therefore many applications like flight simulators, architectural walkthrough or data visualization systems were developed relatively fast. Later on, VR has was applied as a teleoperating and collaborative medium, and of course in the entertainment area.

1.3.2. Data and architectural visualization

For a long time people have been gathering a great amount of various data. The management of megabytes or even gigabytes of information is no easy task. In order to make the full use of it, special visualization techniques were developed. Their goal is to make the data perceptible and easily accessible for humans. Desktop computers equipped with visualization packages and simple interface devices are far from being an optimal solution for data presentation and manipulation. Virtual reality promises a more intuitive way of interaction.

The first attempts to apply VR as a visualization tool were architectural walkthrough systems. The pioneering works in this field were done at the University of North Carolina beginning after year 1986 [Broo86], with the new system generations developed constantly. Many other research groups created impressive applications as well – just to mention the visualization of St. Peter Basilica at the Vatican presented at the Virtual Reality World'95 congress in Stuttgart or commercial Virtual Kitchen design tool. What is so fantastic about VR to make it superior to a standard computer graphics? The feeling of presence and the sense of space in a virtual building, which cannot be reached even by the most realistic still pictures or animations. One can watch it and perceive it under different lighting conditions just like real facilities. One can even walk through non-existent houses – the destroyed ones like e.g., the Frauenkirche in Dresden, or ones not even created yet.

Another discipline where VR is also very useful is scientific visualization. The navigation through the huge amount of data visualized in three-dimensional space is almost as easy as walking. An impressive example of such an application is the Virtual Wind Tunnel developed at the NASA Ames Research Center. Using this program the scientists have the possibility to use a data glove to input and manipulate the streams of virtual smoke in the airflow around a digital model of an airplane or space-shuttle. Moving around (using a BOOM display technology) they can watch and analyze the dynamic behavior of airflow and easily find the areas of instability. The advantages of such a visualization system are convincing – it is clear that using this technology, the design process of complicated shapes of e.g., an aircraft, does not require the building of expensive wooden models any more. It makes the design phase much shorter and cheaper. The success of NASA Ames encouraged the other companies to build similar installations – at Eurographics'95 Volkswagen in cooperation with the German Fraunhofer Institute presented a prototype of a virtual wind tunnel for exploration of airflow around car bodies.

Other disciplines of scientific visualization that have also profited of virtual reality include visualization of chemical molecules, the digital terrain data of Mars surface etc.

Augmented reality offers the enhancement of human perception and was applied as a virtual user's guide to help completing some tasks: from the easy ones like laser printer maintenance to really complex ones like a technician guide in building a wiring harness that forms part of an airplane's electrical system. An other example of augmented reality application was developed at the UNC: its goal was to enhance a doctor's view with ultrasonic vision to enable him/her to gaze directly into the patient's body.

1.3.3. Modeling, designing and planning

In modeling virtual reality offers the possibility of watching in real-time and in real-space what the modeled object will look like. Just a few prominent examples: developed at the Fraunhofer Institute Virtual Design or mentioned already before Virtual Kitchen – for interior designers who can visualize their sketches. They can change colors, textures and positions of objects, observing instantaneously how the whole surrounding would look like.

VR was also successfully applied to the modeling of surfaces. The advantage of this technology is that the user can see and even feel the shaped surface under his/her fingertips. Although these works are pure laboratory experiments, it is to believe that great applications are possible in industry e.g., by constructing or improving car or aircraft body shapes directly in the virtual wind tunnel!

1.3.4. Training and education

The use of flight simulators has a long history and we can consider them as the precursors of today's VR. First such applications were reported in late 1950s, and were constantly improved in many research institutes mainly for the military purposes. Nowadays they are used by many civil companies as well, because they offer lower operating costs than the real aircraft flight training and they are much safer. In other disciplines where training is necessary, simulations have also offered big benefits. Therefore they were prosperously applied for determining the efficiency of virtual reality training of astronauts by performing hazardous tasks in the space. Another applications that allow training of medicine students in performing endosurgery, operations of the eye and of the leg were proposed in recent years. And finally a virtual baseball coach has a big potential to be used in training and in entertainment as well.

One can say that virtual reality established itself in many disciplines of human activities, as a medium that allows easier perception of data or natural phenomena appearance. Therefore the education purposes seem to be the most natural ones. The intuitive presentation of construction rules (virtual Lego-set), visiting a virtual museum, virtual painting studio or virtual music playing are just a few examples of possible applications. And finally thanks to the enhanced user interface with broader input and output channels, VR allows people with disabilities to use computers.

1.3.5. Telepresence and teleoperating

Although the goal of telerobotics is autonomous operation, a supervising human operator is still required in most of cases. Telepresence is a technology that allows people to operate in remote environments by means of VR user interfaces. In many cases this form of remote control is the only possibility: the distant environment may be hazardous to human health or life, and no other technology supports such a high level of dexterity of operation.

The nanomanipulator project shows a different aspect of telepresence – operating in environment, remote in terms of scale. This system that uses a HMD and force-feedback manipulation allows a scientist to see a microscope view, feel and manipulate the surface of the sample. As the same category, the mentioned already before eye surgery system, might be considered: beyond its training capabilities and remote operation, it offers the scaling of movements (by factor 1 to 100) for precise surgery. In fact it may be also called a centimanipulator.

1 . 3 . 6 .   Cooperative   working

Network based, shared virtual environments are likely to ease the collaboration between remote users. The higher bandwidth of information passing may be used for cooperative working. The big potential of applications in this field, has been noticed and multi-user VR becomes the focus of many research programs like NPSNET, AVIARY and others. Although these projects are very promising, their realistic value will be determined in practice.

Some practical applications, however, already do exist – just to mention a collaborative CO-CAD desktop system that enables a group of engineers to work together within a shared virtual workspace. Other significant examples of distributed VR systems are training applications: in inspection of hazardous area by multiple soldiers or in performing complex tasks in open space by astronauts.

1 . 3 . 7 . Entertainment

Constantly decreasing prices and constantly growing power of hardware has finally brought VR to the masses – it has found application in the entertainment. In last years W-Industry has successfully brought to the market networked multi-player game systems. Beside these complicated installations, the market for home entertainment is rapidly expanding. Video game vendors like SEGA and Nintendo sell simple VR games, and there is also an increasing variety of low-cost PC-based VR devices. Prominent examples include the Insidetrak (a simplified PC version of the Polhemus Fastrak), i-glasses! (a low cost see-through HMD) or Mattel PowerGlove.

Virtual reality recently went to Hollywood – Facial Waldo™ and VActor systems developed by SimGraphics allow to "sample any emotion on an actor's face and instantaneously transfer it onto the face of any cartoon character". The application field is enormous: VActor system has been used to create commercial impressive videos with ultra low cost: USD10 a second, where the today's industry standard is USD1,000 a second. Moreover, it may be used in live presentations, and might be also extended to simulate body movements.

2 HUMAN-COMPUTER INTERACTION

2.1 Definition of HCI

Human-Computer Interaction or Human-Machine Interaction is the study that researches the relationship between the system and user. The system can be any type of machines, or it can also be the computerized system and software. This can be seen as a communication language between the computer and human by using a certain interaction to finish the task that designed to be done. The interface of HCI usually means the visible part user can see. User controls on the interface can be small like a home button on the smartphone, big like a panel on the plane craft. The design of a HCI interface should include the understanding of the system to create usability and user-friendliness.

2.2 Detailed interpretation of HCI

The keyword for HCI function is "user friendliness". The HCI function mainly relies on the input, output equipment and related software. The equipment used in HCI are keyboard, mouse, touch screen, speech recognition devices and any other pattern recognition devices. The software that is related to that equipment is the system that offers the HCI function, the use of it is to control the function of the related system and understand every command and request that is sent from the HCI equipment.

At the very early stage, the HCI equipment was the button, like the typewriter. People pressed buttons to create letters. Later, with the birth of the computer, HCI equipment changed to the keyboard and monitor. Operators typed the command on the keyboard, the system ran the process and showed the result on the screen. In the late 1960s, with the rapid development of science technology, more and more people realized the potential of the computer. They thought the computer would become their personal computer and it would no longer be used only by companies or the governments but it could also serve the normal people. At that time although the screen on the computer could only show the green color terminal interface, some people had looked far ahead into the future. One of the important visionaries was an American inventor, Douglas Engelbart. He gave a demonstration on December 9, 1968, of a new computer system called NLS, the oN-Line System, and the prototype of the computer mouse was a three-button palm-sized contraption (Edwards, 2008). This is not only a breakthrough on computer history, but also a big step in HCI development. For the next ten years, many companies started to find out the potential of the computer, the birth of the Graphical User Interface cooperating with the mouse and keyboard, has now become a standard setup for the personal computer. The invention of the computer mouse brought the operation of the computer into a new era.

People are always trying everything new and the same applies to HCI. People want to use the GUI more directly and easily. Another new concept shows up: touch screen, a display that is sensitive to human touch or other kinds of medium. E.A. Johnson is considered as the first one to develop touch screen in 1965 (Johnson, E.A. 1965). So far after half-century of development, people get in touch with touch screens everywhere, ATM machines, PDA, monitors, computers, and phones. After the launch of iPhone in 2007, touch screen technology ushered in a blowout and now we are using it all the time. Touch screen technology is a more natural interaction because the human hand is the most powerful tool in the human history. Using the finger to control the computer that we want to use is a great development in HCI.

With the development of the pattern recognition, the input of speech recognition has been used into the interaction between the operator and computer. This is a big step because people can work on the human natural language, not only follow the binary digits. When we drive a car, we do not need to type on the phone to send messages, we can use our voice to send messages to others, which is really convenient. This can be seen as an intellectualized HCI.

2.3   History of HCI

In 1959, based on how to reduce fatigue when operating the computer, an American scholar Brian Shackel published an article about the ergonomics for computer design which considered the first concept of HCI (Shackel 1959). One year later, Liklider first coined the term Human-Computer Close Symbiosis (Licklider 1960), which is regarded as the enlightenment view of HCI. Later in 1970, two HCI research centers were established, HUSAT at Loughbocough University in the UK, and Palo Alto by Xerox, in America. The next three years there are four monographs related to HCI which gave a direction to HCI development. Then in the next decade, HCI study developed its own theoretical system and categories, on the one hand, independent of ergonomics, and placing more emphasis on the humanities, like cognitive psychology, praxeology and sociology; on the other hand, it expanded from the human-computer interface, underlined interaction from computer to human reaction. Thus the Interface changed to Interaction.

From late 1990s until now, with the rapid development and popularization of the high speed processor, multimedia technology and Internet Web technology, the research has focused on intelligent interaction, multimedia interaction, virtual interaction and Man-machine coordinated interaction, which puts people first. The development of HCI is a change from people adapting to the computer to the computer adapting to people.

2.4   The HCI model

The HCI model is a concept model structure that describes the interactive mechanism in HCI system. So far scientists have come up with many models, like User Model, Interaction Model, HMI Model, Evaluation Model and so on. Those models describe the characteristics of humans and computers in HCI from different angles. Knowing the HCI model is the base of developing a HCI system. It is also a must understand knowledge to design a successful HCI product.

In the model study of HCI, an important model in the early stage is Norman's Model of Interaction which can be considered a logical simplification of an execution-evaluation cycle (Norman, 1988). In this model, Norman divided the HCI process into two stages, execution and evaluation. These stages usually have the following seven steps: Forming the goal; Forming the intention; Specifying an action; Executing the action; Perceiving the state of the world;

Interpreting the state of the world; Evaluating the outcome. Each step is the action of the user. First, the user forms a goal. This is a task for a user. The language of the task needs to be formulated in the different field, but the task might not be that clear, so it needs to be translated into a clearer intention and the actual action to achieve the goal. After execute the action, the user feels the new shape of the system and gives the interpretation of his expectation. At last, if the status of the system reflects the user's goal, the interaction is successful, otherwise the user needs to rebuild a new goal and repeat this cycle. This model is given in Picture 1.

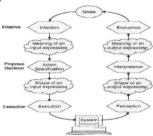

Picture 1. Norman's Model of Interaction

This execution-evaluation cycle model is an effective way to understand HCI. It has a clear thinking and intuition, and it offers a universal frame for the past experience and analysis work. However, it fully pays attention to the user's view of the interaction and only regards the computer system as the interface of interaction and ignore the conduct of the computer system by interface communication. In one sense, it is not a quite completed HCI model.

In 1991, Abowd and Beale improved this model, and made some extension and announced the framework of interaction (Abowd and Beale 1991). This model reflects the characteristics of the user and the system in an interactive system at the same time, which makes the interactive process more complete. This model divided interaction into four main parts: system, user, input and output. Each part has its own language. Besides the user task language and the system core language, it also includes input and output language and these languages express a concept of application area from their own angle. The system language is called the core language which describes computing features in the application area. The user language is called task language and it describes attributes related to the user' intention. This model reflects the general characteristics better in interaction. The interactive process shows information moving through these four parts and its conversion describes the system operation. Input and output together is the human-computer interface. One interactive cycle has four stages: articulation, performance, presentation and observation. The first two stages are responsible for understanding the intention of the user. The next two stages are responsible for the explanation and evaluation of system output. Picture 2 shows the Interaction Framework model.

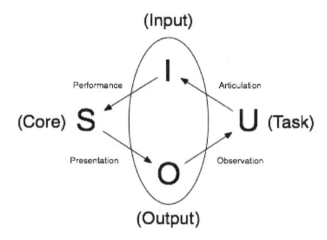

Picture 2. Interaction Framework

15

2.5   Interaction Styles

In fact, the process of HCI is the process of input and output. The user inputs the command through the HCI interface, the computer shows the output result to the user after processing. The form of inputs and outputs between humans and computers are diverse, so are the forms of interaction.

2.5.1   Data interaction

Data interaction is a way that people input data to communicate with the computer. It is one important way of HCI. The general interactive process works like this: first the system sends a prompt to the user and reminds the user of input and how to input; next the user inputs data into the computer using input equipment; then the system will respond to the input and give feedback information, shown on output equipment such as the screen. In the meantime the system will check the input. If it is not correct, it will prompt the user to input again. The data here can be any kind of information symbols, like digit, symbol, color, image and so on. Here are some kinds of data interaction:

Command line: the user types in commands for the program, MS-DOS and UNIX use this style;

Question and answer: the system asks questions, and the user answers. This is simple, but monotonous;

Menus list as much data on the screen that the user can select, like the PDA operation interface and settings on mobile phone;

Form filling: the user types data into the required fields, like the database applications;

Function keys offers special keys, like the computer games;

Graphical direct manipulation: the user can point, click, drag, type. Most Windows systems use this style, known as WIMP which stands for "windows, icons, menus, pointer".

2.5.2   Graphical interaction

Scientific research shows that humans transfer information mainly by speech, text and graphics, Humans absorb more than 70%information through vision, thus the research into graphical interaction is really important and the application areas of graphical interaction are unprecedentedly widespread, like face recognition, handwritten interaction, digital ink, etc.

Graphical interaction, in simple terms, is that, according to the people's behavior, the computer makes, understands and reacts. In this respect, making the computer have visual perception is the first problem to be solved. So far there are three layers of the machine vision system:

Image processing: processing the image to improve the visual effect, transferring the input image to another image that has the required features;

Image recognition: detecting and measuring the specific target in the image. Image recognition gains the objective information to build the description of the image, transfers the image to data;

Image perception is based on image recognition, further researches the interactive relationship in the image and draws comprehension of image content as well as the explanation of original scene, transfers the image to an interpretation.

2.5.3 Speech interaction

Speech is accepted as the most natural, convenient way of information communication. There is 75% of communication in human daily life occurring through speech. The auditory pathway has many advantages. For example, the auditory signal detection speed is faster than the visual signal detection speed; people are sensitive to the change of the voice through the time; providing auditory information and visual information can make people feel a more real and sense of presence. So, the auditory pathway is the most important information pathway in interaction between people and the computer.

Speech interaction is a technology researching how to interact with the computer by using a natural language or a synthetic language. It involves a lot of different subjects, and it not only needs to study speech recognition and speech synthetic, but also research the mechanism of interaction. Generally, there are two methods of speech interaction: one is based on speech recognition and understanding and, mainly relies on the audio signal to interact; the other is speech interaction combined with other interaction styles. In this system, speech interaction is only a part of the interactive system. Some well-known speech interactions are voice assistants in smartphones, like Siri, Google Now and Cortana and people can use them to set an alarm, or open an application, or send a message.

2.5.4 Behavioral interaction

In daily communication, besides using speech interaction, body language is also helps people express their thought. This is called human behavioral interaction. This interaction can not only enhance communication ability, sometimes it can even play a role that speech interaction can not play, like a fashion show, dancing, or sketch comedy, etc. The computer locates and recognizes the user, captures the body language and facial characteristics to understand the human action and behavior and gives a response.

Behavioral interaction brings a brand new interaction style. The computer can predict what the user wants by interpreting user behavior. For example, the computer captures the user's eye to decide if they need a phone call or to surf the Internet. Now on the market, there are some of the experimental projects are trying new forms of interaction (Summers 2013):

Leap Motion is a tiny metallic bar in front of the screen. People can wave their hand to swipe the screen, or rotate image zoom in, or zoom out.

MYO, is like an arm band. It will measure the muscle activity when user uses that arm. Apart from Leap Motion, this is Room Free. There is no need to sit in front of the screen, the user can move to anywhere;

Kinect, Kinect offers a much healthier way of playing the video games. Several motion sensors and cameras will analyze user's motion, so that the player can enjoy the game.

Finally there are three VR/AR devices, Oculus Rift, Google Glass and HoloLens, which will be the focus of the next chapter.

3 VIRTUAL REALITY

3.1   Background

Mobile was the next big thing and although we are in the mobile era, the age of mobile has passed. So we can see large tech companies trying to discover new products that will change the game. These companies do not want to be the next Nokia. The tricky aspect is that no one knows what will be the next. There is only a vision of what can be done to make technology serve humans better. Thus, Google is running a secret lab called Google X which develops some modern or even future tech products, like Google Glass, an augmented reality glass, Project Loon which provides internet Service by balloons, driverless car, etc. Apple and Samsung are investigating wearable devices and smart home. Facebook acquired Oculus Rift to develop virtual reality. All these companies want to take the vantage point in the next round competition and they all believe that virtual reality will mostly be the next thing. If we look back into the past, the computing center changed from the personal computer to mobile, and it not only the size that becomes smaller, the interaction also changes, from the keyboard and mouse to the touch screen. This change makes people get more close to their mobile device. Virtual Reality will bring the user into an immersive environment and it might subvert the entire industry.

3.2   Virtual Reality

Virtual Reality is a computer-simulated three dimensional virtual space, offers virtual sight, sound, touch, and makes the user feel being in a real world, and observe the virtual space with no limit. There is seven different concepts describing virtual reality: simulation, interaction, artificiality, immersion, telepresence, full-body immersion, and network communication (Heim 1993). From a technical point of view, virtual reality has three basic features: Immersion, Interaction, and Imagination. These three "I" features emphasize that the human takes the leading role in the virtual reality system. In the beginning, people could only observe the result from the outside computer system but now they can immerse in an environment that the computer creates. People used to use the keyboard and mouse to communicate with the computer system but now people can use multiple sensors to interact with a multi-information environment. In one word, in the future's virtual system, the aim of human is to let the computer system satisfy people's need, not to force the people make do with the computer system.

3.3   History of Virtual Reality

Virtual Reality is not an advanced technology. In fact, the first prototype of VR device was created in 1968, around half century from now, but it was until now that the word Virtual Reality came into the public view and became a popular topic. This is not strange, after all, at that age, even the computer mouse was just created in 1968, when there was no personal computer, only the computer. Similarly, people might have seen the concept of Virtual Reality as crazy at that time, but this was due to technological limitations.

The prototype The Sword of Damocles was created by the father of VR, Ivan Sutherland. At that time, he only called it as a head-mounted three-dimensional display (Sutherland,1968). This became the first prototype of VR because it defined some key features of VR. First, it has a stereo display. The prototype had two CRT displays to show the image from different optic angles. This is important because it creates stereovision. Then it had a virtual image. created by a computer program which did real-time computing. Next, it had two ways to measure the head position, one was a mechanical linkage, and another was an ultrasonic wave. Next was model generation. In model generation there was a simple cube in the display, but it built up based on the space coordinates. When the user wore this device and turned the head around or changed the visual angle, the cube also changed. Picture 3 shows the system.

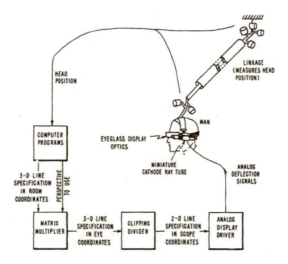

Picture 3. The parts of the three-dimensional display system

We have to admit that, in today's view, The Sword of Damocles is quite simple, and heavy, but was considered as an innovation in 1968 and it was definitely a piece of work much beyond this lifetime.

The most important change in the 1980s was that the parts of a VR device could be purchased individually. We could buy portable LCD displays, the graphics card could support more complicated pictures. A company developed a 6 degrees of freedom head tracking equipment, compared with the past, the accuracy had improved greatly. There were also some gloves that have the joint motion sensors so that the user could have more interaction with the computer. One big moment at that time was that another pioneer of VR, Jaron Lanier, produced the first truly commercial VR product the EyePhone, a virtual reality head-mounted display. After that, VR started to be used in many different fields.

Back to the present, VR technology has a qualitative change than before. The display in the past was not good. So far, the pixel per inch on smartphone has over 500 ppi, brightness and color have also greatly improved. The graphical card can support a large 3D game. The special Inertial Measurement Unit (IMU) used to measure the position of head, it was not only tiny, but also very accuracy. There are also more choices of interaction with the virtual environment, such as, motion control, gesture recognition and speech recognition, etc., we you can find some relatively mature products on the market.

The personal computer, from the prototype to popularization was developed in a few decades, the Internet took a dozen years to connect the whole world, the smartphone and mobile Internet only took less than a decade to take over our daily life. On these grounds, people widely consider that the new interaction on the VR/AR technology will be the next thing.

3.4   Use of Virtual Reality

The use of VR was quite broad. In 1993, according to statistics of VR research projects around the world, there were over 800 projects The result showed that entertainment, education and art take the lead, then is military and aviation, followed by medical science, then robots and commerce.

In entertainment, the rich sensibility and 3D display environment make VR the best tool for developing games. Players want to be in the game, from the very first word game, to two dimensional game, three dimensional game and Internet 3D game. The interaction, the reality and immersion become better and better especially in games like role play, FPS, car racing. VR can fully satisfy the requirements, so now on the market, high tech companies focus on VR games, even on the game platform Steam, there are some games especially designed for VR devices., It is not hard to demonstrate that VR in game play has a bright future.

In the educational field, the interactivity and vitality of VR technology is used to explain abstract concepts in solid geometry, physics, chemistry and geography VR is a powerful tool, and it can save money, without limits of space. In some special training institution such as driving or flight schools, VR can avoid any kind of physical danger. There is no need to worry about a car accident or air crash. In art, VR has a spot sense of participation that can transfer static art like painting or sculpture to dynamic art and people can better understand the thought of the artist. At the same time, VR can enhance the artistic expression; a virtual musician can play many kinds of instruments, so inconvenient people like in the other city can enjoy the concert.

In military and aviation, VR provides vivid simulated training, like the battlefield, or zero gravity environment for astronaut training.

In medical science, VR has a very important purpose. In the virtual environment, we can build a virtual human body, students can understand and observe the organ structure easily which is much more efficient than reading a textbook. In medical schools, students can do autopsy and surgery practice, by using the VR technology, thus eliminating the need to prepare the real body and training expenses. If the VR system is good enough, the advantages are great, surgeons can simulate the surgery in advance, find the best program, and the remote surgery provides some necessary help.

Emergency drills and taking preventive measures are the key in some dangerous industries, firefighting, electric power, oil exploitation, in ensure minimum loss after an accident. Regular emergency exercises are a traditional way, but they cost a lot of money, a mass of manpower, and material resources. If we simulate a man-made accident in a VR environment, this can save money, enhance the frequency of the exercise and ensure the safety. This also works in any other maintenance service. There are more, more fields and more people can benefit from the VR technology.

3.5   Barriers of Virtual Reality

From the above, we can conclude that VR will be a breakthrough in our future life, full of potential, but VR so far is more likely to be used in the commerce or some special fields. There is a long way to go if VR technology goes into the consumer market. There is still a technical limit which is how to provide a true reality environment, and some other problems still do not have a good solution.

Firstly, there is not a real way to step into the virtual world. There is a joke in Oculus Rift developer circles that every time when the staff asks the user to stand up and walk around, the user usually do not dare to move, because at present stage, most of the VR HMD are connected to the computer, this dramatically limits the interactive range. The VR device just covers our eyes, simply changes our eye sight. in our virtual view, it could be a whole world. We can move to anywhere we want, but the current state of VR did not cover all of our sight range. It is also awkward that user still needs to use the mouse and keyboard to control. So in fact we are still sitting in front of the computer. Some developers are try to build a special room for the VR device, but the usual problem is that some of interactions include squatting, hiding, jumping or climbing and these interactions can not be fulfilled.

Second is the input problem, which is the core of interaction. The majority of VR devices can only capture the user's head motion, but not the rest of the body, for example the hand action cannot be simulated. Input is the most important and significant experience for users, otherwise users will confuse where their hands are. Traditional computer peripherals are the keyboard and the mouse, but it is obvious that when the user is covered in the VR environment, it is difficult to use those smoothly. Gamepad seems to be a compromise, but for the traditional PC user, this is still somewhat strange. Some developers try to develop peripherals like a game gun or a sword, but the delayed action is the big problem. In the VR fields, eye tracking and motion tracking is the most efficient way, so development in this area is needed.

Third is lack of uniform standards. There is no doubt that VR technology is still in the primary stage. Each developer has its own way to show the demonstration on a VR device and there is not a uniform standard although developers show much interests in it. As a new platform, the key to success is to attract people's interest and game players are the core audience of VR. but so far game players mainly gather around PCs, home video game consoles, now even mobile devices. So converting people from a mature market to a quite strange market or making these markets compatible is a big problem. Another shortcoming is the lack of talents to develop VR game.

Fourth, it is easier to make people feel tired. Some people may have this experience, when they watch a movie or play a game, when the camera moves fast, it will bring different focus, so human's eye needs to refocus, this might cause some people feel dizzy and sick, especially to female users.

There are others barriers, too. Compared with the first prototype of VR device, the VR device so far is tinier, but it is still quite heavy and not that comfortable. It even looks silly. The price is still high, the consumers will not buy it, and there is a lack of a special platform. Some technology seems to be cool, but it will lose attention as time goes by, so VR has a long way to go. There will be one day when it becomes a new way to change the interaction between human and computer and achieve a more natural interaction.

3.6   Challenges in Virtual Reality

On the market, most of the VR devices are head-mounted display, because HMD can bring user a high level of immersion, but at the same time, this high immersion will cause people to feel dizzy. In fact, this is called motion sickness., The reason cause this problem is that human eyes feel the movement of images which is uncoordinated with the body feeling the movement.

In the human body, feeling the movement of the body is accomplished by relying on the vestibular system. It is like the accelerometer and gyroscope, detect acceleration and angular velocity. When playing a 3D game, like the FPS game, the eyes will tell the brain that our body is moving, but because we are just sitting or standing still, and only use the game pad or keyboard to control the virtual character move, the vestibular system will tell the brain we are not moving, and then the conflict will occur. On the other hand, when a player shakes the head, and suddenly stops, unlike the eyes, the vestibular system will not stop immediately, because of the inertia, it will tell the brain we are still moving, so the signals will be conflicting again.

There are around 30% to 40% people sick of the equipment, 60% to 70% people sick of the content. For the hardware, the resolution of the display screen is not enough, 2K or even 4K resolution might fulfill the need of a HMD, the refresh rate of the display needs to be enhanced, 60 HZ still has space to increase. Recently HTC announced a VR device whose refresh rate is 90 frames per second, and the users said that the problem of dizziness has been solved (Yin-Poole 2015). In the 3D environment, when we move, there is no force feedback, so if some related accessories are added, the problem can be solved. For the software, on the market most of the games have incorporated the software to the VR equipment. The user interface and the play instruction are not suitable for VR devices, so this part should be redesigned. Then, there are two necessary conditions for the frame per second (fps) in the game. First, it should be stable, second, fps must be more than 30, otherwise there will be a problem. If player turns the head from 90 degrees in 1 second, but the fps is not stable and slow for this movement, the brain will think it was done, the eyes will tell it is not done and the player will feel dizzy. Field of View should get close to human eyes, if the FOV is too large, there will be a stretch of the edge of the display, if the FOV too small, when there is movement, the image will change fast. Both of these are not acceptable.

3.7 Oculus Rift

3.7.1 Development history

In 2012, the Oculus Rift project showed up on Kickstarter, a global crowdfunding website. Oculus Rift is a VR HMD designed for game play. It will connect with the VR world to make the player feel real. The target of it on the website was to raise two hundred and fifty thousand dollars, but it seems that people showed a great interest in it. In the end it was funded more than 2.4 million dollars (Kickstarter, 2012). In late 2012, Oculus released the first development kits and after a few months, there were dozens of games supported this VR device: Until now, there are more than 560 apps or games available to use (Riftenabled.com 2015). Early in 2014, Facebook spent around two billion dollars acquire Oculus Rift and from this move, it is not difficult to reach the conclusion that the top tech companies see the huge potential of virtual reality. In July 2014, the second version of the Oculus development kits released, and maybe in 2015, the consumer version will be released.

3.7.2 Oculus Rift teardown

The development kits 2 include a headset, several cables, one camera, two lenses and several different plugs for different countries. The motherboard has many sensors, like the accelerometer, gyroscope and magnetometer, it also has a built in latency tester, IR sensor, the camera is used to capture the IR sensor and the display is 5.7 inches, per eye with a resolution of 960 X 1080 (iFixit, 2014). Picture 4 shows the Oculus Rift DK 2.

Picture 4. Oculus Rift DK 2

3.7.3 How the Oculus Rift works

Basically, Oculus Rift is an external monitor supported HDMI and DVI input, in the meantime, it is a USB device that can detect head motion. In order to achieve 3D image, the screen displays two images, the left eye sees the left half of the screen, and the right eye sees the right half of the screen. In development kit 1, the resolution of screen is not that high, so in development kit 2, Oculus increased the resolution to Full HD and the user gets a better experience. A set of lenses are placed on the top of the screen, and this is the key to fulfill the stereoscopic 3D image which is used to focus and reshape the picture. The reason why Oculus Rift is the star in VR devices is that it can nearly restore the human's field of view, almost 180 degrees forward facing horizontal, 135 degrees vertical (Wikipedia, 2015), Oculus Rift has a 100-degree vertical FOV (Oculus VR, 2014) which enhances the immersion of user. Two convex lenses are the secret, based on the simple optical theory, which magnify the screen. As shown in Picture 5, by using the convex lens, it can magnify the original length from to '. However, there is a consequence, at the same time the convex lens magnifies the image, the distortions will exist, so on software layout, Oculus Rift gives a solution to correct the distortions by creating the same size but opposite distortion. Thus the man-made distortion can counteract the optical distortion and the user can see the normal image through the screen.

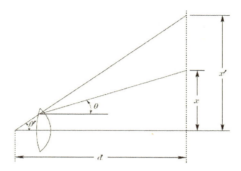

Picture 5. Magnifying glass theory

Head tracking is important in VR devices because in the real world, people can turn their head around to see the world. In the first prototype of VR device, the mechanical linkage was used for head tracking, nowadays thanks for the technology development, it can be replaced by much tinier sensors, gyroscope, accelerometer and magnetometer. In the second development kits of Oculus Rift, one more external camera helps to track the headset position more precisely. First, the sensors create a 3D position vector for which the coordinate system uses these axis definitions: X is positive to the right, Y is positive in the up direction and Z is positive heading backwards. The opposite directions in each axis have the following rotations: Pitch is rotation around X, positive when pitching up, Yaw is rotation around Y, positive when turning left and Roll is rotation around Z, positive when tilting to the left in the XY plane (OCULUS VR, LLC, 2014). Picture 6 shows the head tracking sensors system.

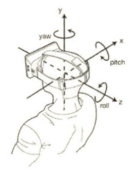

Picture 6. The Rift coordinate system

In the first development kits, there is no IR camera, which means that it can track head rotation very well. However, when we move our whole body, such as when sitting or jumping and swinging, the virtual head on display would not move. This decreased the immersion and reality. Therefore in the second development kits, some changes were made and an IR camera was added. On the headset there are several IR LEDs. Capturing the set of IR LED opens new ways of game play, is more comfortable and more immersive. Players can hide in a corner, swing their body to dodge bullet, and much more. Picture 7 is the position tacking system.

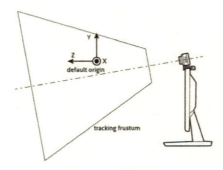

Picture 7. Position tracking camera

3.7.4 How to solve the latency

To enjoy the game, players can bear many things. 3D games might make some people feel sick and VR devices might aggravate this feeling. The feeling of dizziness that Oculus Rift causes comes from the latency of the whole system. The sensors, the screen technology, the GPU speed are all the factors cause the problem. In general, latency occurs when user moves his head, and the image shows the right one on the screen, there is a time sensors will make response, image transmission and display response. Even with today's technology, latency can not be eliminated completely. It can only be controlled to be as low as possible. With the power of the processor, on the market some VR HMD decrease latency to 100 millisecond or 40 millisecond, but it is not enough, research shows that 20ms is the watershed, 15ms will remove the simulator sickness (Abrash, 2012).

Oculus Rift divided the latency time into six parts:

- The user inputs information,

- The USB cable transforms the command from Rift to computer,

- The game engine translates command to GPU,

- The CPU gives order to write a new image,

- The display transforms pixels to an image,

- The image is fully generated.

Oculus is trying hard to cut down every step's time. The first step is to minimize the input latency which means finishing the transfer from head motion to digital command as fast as possible. Differently from the prototype of the Oculus Rift, now the current version has a built-in tracker integrated with gyroscope, accelerometer and magnetometer and time has decreased from 15ms to 1ms. Then the signal transfer from Rift to computer will make 1 to 2ms latency and the problem is with the cable, so unless the USB cable is reinvented, we can not reduce the latency time.

Next, game developers will take responsibility for the latency The game fps plays a particularly important role. At present, most games are around 60fps, which means that every image transfer to GPU will cost about 16.67ms, so if developers improve the speed 1 times, the latency will be the half as before. The GPU transfers the command to pixel on display through USB cable. Some pixel transform fast, for example, black to white only takes less than 10ms, some pixels need more time. To save the time, every single pixel starts to transform once it has received the command from GPU. The Rift begins to write from bottom to top, so when the command transfer to the top, the pixels at the bottom will have finished the transformation. The whole process takes 20 to 30ms, based on a 60fps game, the whole latency takes 40ms which is still too much for a VR device.

It is seems that latency time is mainly wasted on the screen. Oculus Rift used to use LCD display, but OLED technology can have an extremely fast refresh rate, lasting only a single millisecond. So using OLED display might solve the problem. However there is only one manufacturer offering OLED technology, that is Samsung, but Samsung has not sold that technology to any third party. However, on second development kits, the display use OLED technology. From the teardown of Oculus Rift can see that it just uses the Galaxy Note 3 display panel directly on the screen, a 5.7 inches Super AMOLED 1080p screen, and Oculus slightly overclocks the display panel, and changes the refresh rate from 60Hz to 75 Hz to decrease the latency time on screen. Besides that, the internal tracker has a really high data sampling capability, 1000 times per second, which not only reduces the latency, but also has a capability to predict the head position that user may move to. If the head movement is fast, it will not be able to stop immediately as it can only slow down from one point. Thus if players move their head, at one period of time, the tracker can predict the next position of the head, after the head slows down, the prediction moves down to 0. This process not truly changes the latency time, but it allows players see the image much faster. This can change the latency subjectivity for about ten milliseconds.

All those efforts have successfully shortened the latency into 20ms, which makes the Oculus Rift the best VR HMD so far, except latency. However there are still other things that need to be resolved before the product is delivered to the consumer, like the accuracy of the tracking, the quality of the image, the image resolution. However with the support of Facebook, the enthusiasm of the developers and the development of the technology, we can hope that Oculus Rift will offer the best VR experiences to the player one day.

4 AUGMENTED REALITY

4.1   Definition

Augmented Reality is an advanced technology based on Virtual Reality, Augmented Reality is a technology that calculates the position and angle of the camera and adds related images on to the relevant real world in real time. The goal of this technology is to put the virtual world and real world on screen together and make interaction. The concept of the Augmented Reality was first mentioned in the 1990s, Tom Caudell and his colleagues from Boeing designed a system that overlaid some text tips and routing paths on machinist's head-mounted display in real time. The information could help the machinist teardown the machine easily (Lee, 2012). So far there are two general definitions of Augmented Reality, the first one is Milgram's Reality-Virtuality Continuum mentioned by Paul Milgran and Fumio Kishino in 1994 (Milgram and Kishino 1994). They point out that real environment and virtual environment on two sides of a line, as Picture 8 shows below, the middle space between them is called Mixed Reality, the space near the real environment is called Augmented Reality and the other side is called Augmented Virtuality.

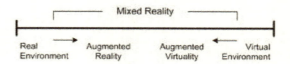

Picture 8. Milgram's Reality-Virtuality Continuum

The other definition is more well-known. According to Ronald Azuma (Azuma, 1997), Augmented Reality contains the following three points: It combines real and virtual; it is interactive in real time; and it is registered in 3D.

Augmented Reality is in a way combined virtual technology to re-observe the real world. AR is able to provide information we cannot gain from the real world, at a deeper level. This information can make the world more diverse. A strange street, in normal eyes, is just a street, but from the sight of AR, it will cover a different 'information coat', or another so called ' The Long Tail'. We will know what the name of this street is, what kind of shops it has, which are some of the best shops. People can know everything they want. These are already some of the characteristics in some AR product at present. With the development of technology, AR could be more advanced and makes the interaction expand from accurate location to the whole real environment, from the simple communication between human and screen to integrating humans into the surrounding environment.

4.2   Types of AR systems

A complete AR system is combined with the real-time hardware components and relevant software. There are three main types of AR systems:

Monitor-based AR system: It is based on the computer monitor, a camera will take the image from the real world, input it to the computer, then compound the real world image and virtual image simulated by the computer graphics system, finally it shows on the screen and the user will see the final AR image. It is simple, but it cannot give the user much immersion. Picture 9 shows how the monitor-based system works.

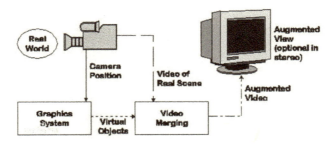

Picture 9. Monitor-based system

Optical See-through AR system: It is based on the optical principles and uses a head-mounted display. Unlike the Monitor-based system, the user can see the real world in this system. There are optical combiners in front of the user's eyes, which can be imagined as a glass, so the real world can be seen. The head tracker will track user's head position, then the scene generator will generate related images and show them on the monitors and the user can see them on the optical combiners. Picture 10 shows how the optical see-through system works.

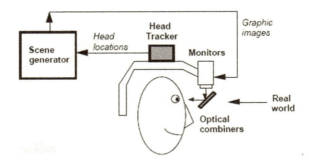

Picture 10. Optical see-through system

The Video see-through AR system, also needs HMD, and it is based on the video synthetic technology, quite similar to the optical see-through system. The only difference is that user sees the real world through the monitor. Video cameras will help to record the real world information before eyes, the video of real world will be combined with the virtual images and then the combined video will be shown on the monitor. Picture 11 shows how the video see-through system works.

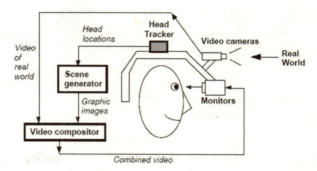

Picture 11. Video See-through system

To compare optical and video see-through systems, they both have advantages and shortcomings. This is a short analysis of them. The optical system is simple, because the system only needs to process one virtual image, the real image just directly seen by user's eyes, when the head tracker tracks the head position and shows the relevant images. the latency is short. However, the video system needs more time to handle both real and virtual images, so the delay is longer than in the optical system. Besides that, because of the extra video cameras, the video system is heavier than that of the optical system, and because the video compositor needs to process and combine the real and the virtual images, so it is more complicated than the optical system.

The resolution used to be a problem of the video system in the past, because the display technology was not advanced enough and human eyes can see the pixels clearly on the screen. However, now with the development of the screen technology, the PPI of the screen has developed far beyond the limit of human eyes, so this is no longer a problem. Another shortcoming that related to eyes is the eye offset. In other words, the video camera is not exactly at the same position as human eyes and might be a little higher on the HMD, so this might cause some misjudgments of the real environment. For example, if an engineer is trying to maintain a machine, the misjudgment of the position will most likely create a mistake. The solution to this problem is, use the mirrors to change the path of the light which will make the system slightly more complicated.

The optical system has a vital drawback. In order to reflect the information on the monitor at the optical combiners, the design of the combiners cannot be fully transparent, so the amount of the light will be reduced and the users may feel they wear a pair of sunglasses. On the other hand, the video system does not have such a problem.

One of the greatest strengths of the video system is that it has a wide FOV. The vertical range of human eyes is about 135 degrees, but the video camera can have 360 degrees of view if there are several cameras working at the same time. The optical might occurs the distortion so that the image will be rather unreal and strange, but with the video camera, the real world image is digital, by using the software, we can correct the distortion, and finally an undistorted image can be displayed on the screen, and the user will have a better experience.

4.3   Use of Augmented Reality

The use of Augmented Reality has many application fields similarly to the Virtual Reality, such as sophisticated weapons, the development of aircrafts, virtual training, and entertainment. Because it has the characteristic of augmenting the real world, it has applications in fields like medicine, the maintenance of precise instruments, military and engineering. AR technology has more obvious strengths than VR technology.

Medical field:

Application in the medical field offers huge benefits. As mentioned in Chapter 3, the VR technology can be used in this field. Students can learn the human body and do the surgery. However AR technology can do better, students still need to observe the true human body and surgeons need to have an operation in real life. At this time, AR technology can play a supporting role, for example it can indicate which places on a human body should start use the scalpel to do the incision or which place has the tumor. This assistance can help the surgeon perform his surgery more precisely and enhance the success rate of the operation.

Military field:

AR technology can help the troops obtain location in real time, and the geographic data on a strange battlefield. AR also provides basic navigation and airplane information, target location in the military aircrafts.

Museums and historic sites:

When we visiting a museum, we can see many cultural relics, some of those relics are not complete, and if there are many visitors, it is hard to see the explanation of that relic, some of the outdoor historic sites also fragmentary. AR technology can bring a whole new experience, visitors can see the reconstruction of the imperfect relics, what its original shape was.

Entertainment:

AR games can improve the interaction between players and the real environment. The mobile game Ingress is probably the best example of the AR games. This game has a simple science fiction background. There are two camps that player should choose from at first, then player needs to find and collect the items on the map, the item can be seen as the experience, then find the 'Portal ', attack or defend. The map in the game is based on the Google Maps, so the location of the item and the Portal is real, they can be a museum or a specific building. The player needs to go to that place in the real world to collect the item and get close to the Portal. This is a worldwide game and there are some clubs offline gather the player to hold some events, play this game together, so this is a really cool thing and much more interactive with the player. Picture 12 is the user game interface of Ingress, and, the map in the background is of New York City.

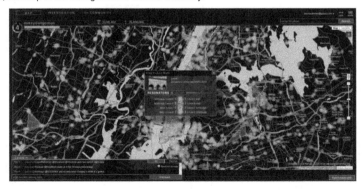

Picture 12. Ingress game

Navigation:

The traditional way of navigation on the mobile phone is good, but not interactive, for people in a strange place, sometimes cannot be sure if this is the right place shown on the map. Using AR technology in t navigation, those problems can be solved perfectly. Nokia City Lens is the best software to describe this. Through the camera and GPS on a mobile phone, the camera

TURKU UNIVERSITY OF APPLIED SCIENCES book | Xing Huang can display the scene around you, like shops, restaurants, hotels, even bus stops. There are several labels on the screen showing the exact location of those buildings, and their approximate distance. When we click on one of the labels, it will show the detailed information, like the name of the restaurant, the special food of it and the feedback from people. When the place is a little far from your location, the software can give turn-by-turn navigation and ways of transportation to arrive at the destination. This is a powerful tool when travelling to an unfamiliar place, more convenient and interactive with the people. Picture 13 below is the user interface of this software.

Picture 13. Nokia City Lens

Other fields:

In industrial maintenance, the HMD can offer supplementary information to the user, including the virtual panel, the internal structure of the device and the blueprint of the item. In the television relay, such as basketball games, it can overlay the basketball player's data, like points, assists, rebounds, so the audience can have more information. In video chat communication, the user can wear some funny virtual hat or glasses to make the video chat more interesting.

4.4 Realistic way to achieve AR

There are about three ways to achieve AR: GPS and sensor, marker recognition and PTAM.

GPS and sensor are mostly used on the smartphone. At present, nearly all the smartphones have those sensors The GPS sensor can get the longitude and latitude of the location and some of the smartphone even have a barometer sensor to get the height, the electronic compass to get the orientation. The accelerometer sensor can get the tile angle and lastly, using the camera on the phone to get the surrounding image and the related information can be shown on the screen. This is the simple way to achieve AR technology. Nokia City Lens is a good representive of an AR application. Therefore, we can conclude the AR is not a high-end technology that normal people can not used. On the contrary AR technology is already in our daily life.

Marker recognition stores the Marker images in advance, uses the camera to capture the image, then uses the image recognition technology to recognize the Marker image in the database, also gets the position and direction of the Marker, finally overlays related information and displays it on screen. This technology is based on an open source project called ARToolKit. It is a computer library written in C/C++ language and helps developers to create an AR application more easily. In order to develop an AR application, the most difficult part is to overlay the virtual images on the real images in real time and make a precise alignment to the real world. ARToolKit uses computer graphic technology to calculate the relative position between the camera and Marker, so developers can overlay the virtual image on the Marker successfully. ARToolKit includes tracking library and full source code and developers can adjust the interface in different platforms.

One example of Marker recognition is a mobile application called Blippar. Blippar cooperates with some famous brands to help them show user friendly advertisements. When we open the application on the smartphone, the application will scan the Marker 'B'. If there is no 'B', it is still ok if the advertisement has cooperation with Blippar, after a few seconds, an AR advertisement will show on the screen and user can make a simple interaction with the advertisement, can click on it to see more information about the product. Nevertheless, this way to achieve AR has many limitations, because it should define the Marker beforehand, and it cannot gain information outside the Marker, so it is only suitable for games, books and advertisements. Picture 14 is how Blippar works.

Picture 14. Blippar application

Image analysis and recognition uses the camera to analyse the objects, space and the surrounding environment and then overlays related information. The most famous related project is Parallel Tracking and Mapping, shortened as PTAM. It is a camera tracking system for AR, first showed on ISMAR 2007. Georg Klein and David Murray from the University of Oxford published a book called Parallel Tracking and Mapping for Small AR Workspaces (Klein and Murray, 2007), it is an open source project whose main purpose is to capture the feature points using the camera, then detect a plane, next build a virtual 3D coordinates on that plane, and create the composite image and CG. The unique feature is parallel processing which is used to detect the 3D plane and create composite the images, so it can save a lot of time to solve some large and complicated operation. This could be one method to achieve AR, because there is no need to find the special Marker. AR can be achieved by analyzing the image and building real images on coordinates and compose AR more naturally. However this method also has some problems, like the large number of data to be processed, differentiated and analyzed on the plane and the vertical plane. The technical difficulty of PTAM is relatively large, so PTAM AR still needs to be developed. Picture 15 shows how PTAM works.

Picture 15. PTAM

4.5   Challenges in Augmented Reality

The first challenge is the same as in VR. In order to fully immerse the user, it is best to have the image in real time, but with different problems. Latency is always a problem because AR has a high demand for real time synchronization. As the user moves in the real world, the virtual information should be overlaid as soon as possible, so the user can get the correct information. Minimized latency will enhance the immersion of the user.

Then there is the computing capability. Unlike the VR which only needs to handle the virtual signal, AR has to analyze the real world environment and at the same time manage the virtual signal. There is vast amount of data in AR, so the hardware should be powerful. The hardware is a head-mounted device, so it should be tiny, light, at the same time. The power should be as low as possible, otherwise the battery life and the heat will confuse the user. Based on Moore's law, a small but powerful hardware is not a problem in the future. The capture and tracking system need work accuracy, so that they can give the user the right information, the software should also be optimized to filter the information, remove useless data and keep the useful data.

Another challenge that needs to be addressed is the upstream bandwidth. In order to show the virtual information on the real world, the AR device needs to send the real world data onto the nearby server, and the nearby server will give back the information. Because of the character of the AR technology, there is a high demand for the upstream bandwidth, but nowadays the wireless internet device has a large downstream bandwidth. According to a recent study, shown in Picture 16 below, the downstream bandwidth is 3 to 10 times larger than the upstream bandwidth (Broadband Backhaul, 2014). So for purpose of better use of AR HMD in the future, designing a new wireless Internet infrastructure is necessary.

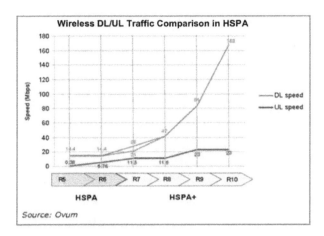

Picture 16. Wireless DL/UL traffic comparison

The last but not the least concern of AR device is privacy. AR devices are very attractive and, a lot of people want to use them, but the privacy still needs to be considered. Like Google Glass, the camera on it allows it to take pictures or videos without being noticed, which means that a user can record a movie at the cinema, and when walking on the street, the AR technology might find out the users name, age and other information through liking in Facebook. Although these actions looks trendy, they actually violate privacy and that is why after Google Glass was launched, some companies forbade their employees wearing Google Glass on some special occasion. So these kinds of devices need a serious discussion on privacy.

4.6   Google Glass

4.6.1   Brief introduction

Google Glass, first presented in 2012 by Google Glass (Goldman, 2012), was seen as the first truly AR hardware device in the world. The aim was to offer ubiquitous computing to consumers, Google Glass has some functions that the smartphone has, like taking pictures, having a video chat, sending messages, surfing the internet and giving a road direction. So far it has two explorer editions at the price of 1500 dollars, but in January 2015, Google announced that it discontinued the explorer edition of Google Glass. The public has no idea if this move means that Google gave up this project or not. Although Google Glass seems dead, it brings several labels into `the public eyes: Augmented Reality, head-mounted display, wearable device and wearable computer. Augmented Reality has become a hot topic in the last two years and people start to look into new ways of interaction with the computer.

4.6.2   Composition of Google Glass

The main components of Google Glass are shown in Picture 17 below. It has five megapixels camera, can record a 720p video, on the side of the glass there is a track pad, the user can swipe forward, backward, down and tap. Inside the trackpad is the motherboard, with CPU, gyroscope, accelerometer, compass and some sensors. There is a one-day battery on the back, the sound system uses the bone conduction transducer to play the sound, and Bluetooth is used to connect to the mobile phone. At the front, there is a display, the size of it is like a 25-inch high definition screen from eight feet away, and there is a prism used to reflect the visual images from the projector to the user 's eyes (Support.google.com, 2015). The weight is so light so it just likes normal glasses.

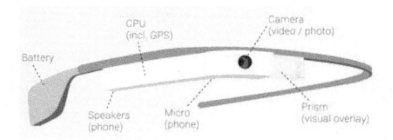

Picture 17. Composition of Google Glass

4.6.3 How Google Glass Works

Here is a simple explanation on the display technology behind the Glass, how it shows virtual information to user's eyes. There is a projector on the Glass. Unlike the normal projector technology, Google Glass uses Liquid Crystal on Silicon, a display based on reflection which has a better light utilization, so it can achieve a high resolution and color performance, and it is power saving and cheap, thus it is the best choice for the mobile projector. First, the LED light shines through the polarizing beam splitter onto the LCoS panel, then it reflects on the PBS again and refracts at a 45-degree angle to the collimator at the end of the prism, next reflects on the partially reflecting mirror and refracts at another 45-degree angle into user's eye. That is how the user sees the virtual overlaid image. Depending on how the user wears the Google Glass, normally, the virtual image will show on the top right corner of the visual field, and, at the same time, the real world lights just shine through the partially reflecting mirror into the user's eye (Google Inc., 2013). Besides the display, the communication between the user and the computer is also important in the AR device. Voice communication is one of the interactions on Google Glass, but there is a problem, if we want to use Google Glass outside in a noisy environment, to keep the voice communication clear, Google uses the bone conduction transducer. Bone conduction uses human cranium as the medium of sound, and conducts the sound directly into the human brain, so it has the best effect of sound conduction, and s the Glass more convenient. Picture 18 below shows how Google Glass display works.

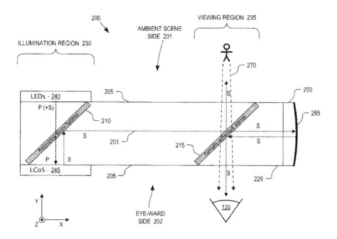

Picture 18. Google Glass Explorer Optics Schematic

4.6.4 Features of Google Glass

Google Glass has the features that a smartphone also has. It runs the customized Android operating system, so it is just like a wearable smartphone. Voice recognition is one of the main interactions of Google Glass. When the user says "OK, Glass", a menu will show with some icons indicating actions such as take pictures, record a video, use Google map, call somebody, send messages or e-mail. This is very convenient when the user does not have a free hand to hold the phone. For example, if we want to take a picture, we just need to say "OK, Glass, take a picture", and Google Glass will take a picture, or there is a picture button on the top of the glass and we can tap that button to take a picture. The same applies to recording a video. This is one of the cool features which make Google Glass special because the picture can be taken as a first person looking, the special camera angle can record your personal life in a unique way and we can have both of the hands free.

Another killer application is the Google Map on the glass. HMD is the best device to show the map application to the user, because it is hand free. Thus, it does not matter if the user is walking, riding or driving, the map application on

HMD is extremely convenient. When wearing a Google Glass, we just need to say "OK Google, get the direction to some place", and we can choose the way we go there, the turn by turn navigation will show on the right side of the eye, so when we are riding a bike, we do not have to worry about accidents because both hands are on the wheel.

Other basic build-in features include sending or receiving e-mail and text messages, Google search, looking at the weather. Google Glass has the potential to become more functional and powerful if the developers are willing to develop new applications on this platform. Many third party developers and companies have already made some applications like some news apps, The New York Times, social network apps like Facebook and Twitter, note apps like Evernote, and facial recognition, exercise, translation.

In conclusion, Google Glass is a small but powerful device. It includes a lot of high-tech hardware like Bluetooth, Wi-Fi, gyroscope, accelerometer, compass, camera, microphone, speaker, and a tiny screen. The shape is just one pair of normal glasses. It uses the most natural way of human communication: voice recognition to interact with the computer, all we need to say is "OK, Glass". The convenient navigation will lead us to our destination, we will never get lost. Real-time acquisition is like an intelligent assistant, there is no need to worry about forgetting important things, because it can remind the user the flight information, hotel information, meeting time. It also supports both Android and iOS operating system, works as an extra device, answers phone calls. The design of the Google Glass is fashionable and it even appears in some fashion shows. This is what the future technology looks like and is the core of human computer interaction, ubiquitous computing. The computer will integrate into the surrounding environment, disappear from the eyesight, people can gain and manage the information in any place, any time, any way.

4.6.5 Main problems of Google Glass

It is a fact that Google has officially announced that it will stop selling the explorer edition of Google Glass. After around two years, here are some possible main problems of Google Glass.

The first one is the privacy. It is not only Google Glass that has this problem. All the wearable smart devices have the same concern, especially those devices that have cameras can take pictures or record videos without any notice. Because of that, Google Glass has been forbidden in many public places, like bars and restaurants. At the meantime, because of the pirated movies, a lot of cinemas in America also prohibit Google Glass and other wearable video equipment.

Second is the safety and security problem. When wearing the Google Glass, due to the virtual image on the top right corner of the visual sight, in order to see the information, the user needs to focus on the corner. This might cause distraction, especially for the drivers, and accidents. On the other hand, Google Glass is connected to the mobile phone through Bluetooth, so hackers can steal the smartphone password by cracking the Glass, and steal other information inside the mobile phone. However there are other ways to improve security, for example by using fingerprints to unlock the phone.

The third problem is the lack of "killer application" and the target audience is not clear. So far on Google Glass, some of the core applications just work like on the mobile phone. There is no application that motivates the users to use Google Glass to achieve that aim. Google Glass acts too much like a phone on the head. It becomes an unnecessary product unless there is one or several "killer applications" that makes the user want to keep wearing the Google Glass. Except that, the market positioning of Google Glass is unclear. Google wants Google Glass to be a necessary item in our normal life too much. Just like the shape of Google Glass too much like as a normal pair of glasses, this strategy is not good for a potential game changer product. At present, the smartphone is still the main market and people can now do anything on the smartphone, thus it is better for Google Glass to be used in specific fields.

As the first true augmented reality wearable device, Google Glass takes too much pressure. Although it is no longer on the market for now, it is hard to say whether Google Glass is success or failure. Google Glass is a quite advanced product, there is a long way for it to become popularized, but it represents the thought of the next way of computing, a much more interactive product for humans. It is undeniable that Google Glass has many flaws, but with the saturation of the mobile computing market, it needs a new product to step into the next revolution. Google Glass is more like a symbol that bring AR technology into the public eye, encourages more and more companies to develop this technology together, and makes AR become the next stage of human-computer interaction.

4.7 HoloLens

4.7.1 Announcement of HoloLens

On the 22nd of January 2015, Microsoft announced a fantastic Augmented Reality HMD called HoloLens (Hempel 2015), a short name of Hologram Lens. A lot of people became shocked by watching the demonstration which felt like moving scenes in science fiction movies to the real world. Because HoloLens is an Augmented Reality device, there is a short comparison to the Google Glass. Unlike the Google Glass which can only overlay a two-dimensional layer for the user, from the demonstration of HoloLens, we can see that it can also overlay three dimensional images, just like the same as hologram technology, so the user will feel more vivid by using HoloLens. From this angle it is more powerful than Google Glass. Then the HoloLens supports gestures, which means that the user can tap, swipe, pinch, zoom in and out through the air. This greatly strengthens the interaction between user and computer. Last but not the least, in our opinion, HoloLens will not have the same end as Google Glass because Microsoft Pictures out that this kind of technology should first be used in inside room, like the home and office, so people will not feel strange that one guy wears a big HMD and make some weird gestures. Afterwards this behaviour will be accepted and popularized by people and it is not hard to see people wear it outside. Picture 19 below is the prototype of Microsoft HoloLens.

Picture 19. Microsoft HoloLens

4.7.2 How Microsoft HoloLens works

At present, Microsoft just announced HoloLens but it is still needs time for the development version, there are no technical details from the official announcement so far, so the next points is only represent our point of view that HoloLens probably works like this. First, although Microsoft called it a Mixed Reality Product, HoloLens is basically an Augmented Reality product. It has an independent computing unit, CPU and GPU from Intel and one Holographic Processing Unit (HPU) and it is an application-specific integrated circuit customized by Microsoft.

It is clear that HoloLens is not a VR device, totally different from Oculus Rift whose feature is to immerse the user into the computer simulated three dimensional world but it cannot help people better understand the real world. HoloLens is not Google Glass. It is much more advanced than Google Glass, HoloLens has the ability of three-dimensional perception, after scanning, HoloLens can model the surrounding world, but Google Glass can only see the two -dimensional pixel value. Next HoloLens is capable of 3D rendering and stronger interaction, not only voice recognition but it can also use gestures to control the device. HoloLens is not an AR device based on camera for mobile phon. As mentioned before, these applications can only detect the panel of the special image, but HoloLens can detect 3D scenes from any angle.

Going back to the definition of Augmented Reality, in order to fulfill the AR, it must first understand the reality through the camera data to gain the 3D depth information. So why can the camera on HoloLens perceive the depth? It is all because of the technology accumulation on another Microsoft product, Kinect. There is a CMOS infrared sensor on Kinect and also on HoloLens. There might be 4 cameras two on each side, and those four cameras with the sensors can cover a wide horizontal and vertical angle. The CMOS infrared sensor can detect surroundings by black and white spectrum. Pure black means infinitely far, pure white means infinitely near and the grey zone between them means the distance between the item and sensor. The sensor collects every single point in the visual field, simulates a depth map in real time. The depth map is used for Stereo Vision technology. The aim of Stereo Vision is to rebuilt 3D scenes by using the camera, usually two cameras to gain the distance data. Unlike the human eyes, the computer stereo vision is quite backward. One of the main challenges of AR on the HMD is to better manage the stereo vision. Picture 20 is the sketch map of the Stereo Vision.

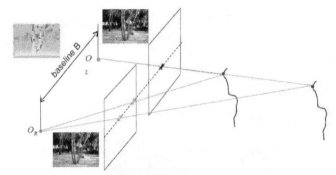

Picture 20. Stereo Vision technology

The first step of the Stereo Vision is to correct camera distortion. The lens on the camera has the distortion, so in order to gain the precise data, the camera needs to be corrected., The common way of correction is to take a few pictures of the chessboard, then calculate the matrix parameter of the camera and the distortion can be corrected. The next step is image rectification. Because two cameras stand in different positions, there is a deviation of two cameras. The left camera can see the leftmost image and the right camera can see the rightmost image. The aim of image rectification is to get the same part of the image. Next is image correspondence. Left and right images should be matched and the last step is to use the reprojectImageTo3D function in OpenCV to build the depth map (Docs.opencv.org, 2014). However, getting only one depth map is not enough. It only represents the image at one moment. To get the whole 3D scenes, there is a series of depth map that need to be analyzed. Picture 21 shows the depth map.

Picture 21. Depth map

4.7.3 How to rebuilt 3D scene from depth map

The answer to rebuilding a 3D scene is SLAM, Simultaneous Localization And Mapping. It is a system used in robots, driverless cars, unmanned aerial vehicles and some other computing systems. The aim is that the robot starts in an unknown environment, using the SLAM system to locate its position and rebuilt the 3D map in real time, at last arrive at the destination. The SLAM system deals with a very philosophical problem, what the world looks like and where I am. Based on SLAM, Microsoft created its own algorithm called KinectFushion (Izadi et al. 2011) and published two theses. After using SLAM on Kinect, there is a high possibility of SLAM used on HoloLens. The user wears HoloLens and moves in a room, the system will gain different angles of depth maps, iterate in real time and accumulate those depth maps to calculate an accurate 3D model in a room.

KinectFushion rebuilds a 3D scene from the depth map in four main stages. The first stage is called depth map conversion. It converts the depth map from the image into space coordinates, saves them as floating points and calculates the vertex coordinate and normal vector on the space. The second stage is called camera tracking and it calculates the camera pose, including positions and orientation under global coordinates, by using the iterative alignment algorithm to follow the values, thus the system will always know the change of the camera from the original position. The third stage is called volumetric integration and it integrates the depth data from stage two into a 3D space. The fundamental element is not the triangle but square. In the demonstration of HoloLens, there is one scene that a user wears HoloLens in his room to play the MineCraft game. Thus, some of the HoloLens applications run on stage three. The fourth stage is called ray casting and it is a common way of 3D rendering, because HoloLens is AR HMD, the screen is transparent, so there is no need to render the room because the human eye has already rendered it. Picture 22 depicts the stages of KinectFushion.

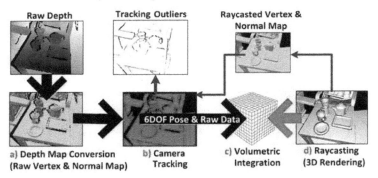

Picture 22. KinectFushion

4.7.4 How gesture recognition works on HoloLens

There is also that HoloLens will use the same gesture recognition technology on Kinect. Gesture is one part of communication in human daily life, use this technology on computer, people can free their hand, use a more natural way to interact with the computer. In the field of HCI, gestures represent a simple command, have arbitrary meanings, are based on conventions and might be misunderstood. To achieve gesture recognition, Microsoft created a Skeleton Tracking system.

On Kinect, the Skeleton Tracking system will analyse the depth map data to extract the human body silhouette, then create the coordinates of human joint. There are around 20 joints on the human body that need to be tracked, and these 20 joints simulate a skeleton system. By recording and comparing the skeleton with the database, the computer does the related task.

As on HoloLens, the camera only needs to capture the hand gestures and does the same thing to achieve the gesture recognition. However, as mentioned before, unlike the command, the gestures cannot be exactly the same every time, so this kind of gesture recognition highly relies on the machine learning. The recognition engine will record, handle and reuse the data, so as the time goes by, the precision of gesture recognition will be enhanced.

4.7.5 Strengths and drawbacks of HoloLens

From the demonstration of HoloLens, this AR HMD can be used in many different scenes. First is the game play, users can play a game like Minecraft based on their real living room, some buildings on the sofa, some on the table and some on the floor. It is a quite interesting experiences. The second demonstration is the video chat, by using Skype, a father can help his daughter fix the tap, the father can teach and show the virtual array on how to operate, so the daughter can fix the problem easily. The third demonstration is in the office, an architect is designing the shape of a fancy motorcycle, she can see the 3D model of the motorcycle and adjust a little as she wants, this is a much more direct way that architect can change his design. The final demonstration is cooperating with NASA to explore the land on Mars, user will be on the surface of the Mars, the color, the landform, just like the real.

From the four demonstrations HoloLens, we can see that it has great potential to be the game changer as it can be used in many different fields. Compare with another AR HMD Google Glass, HoloLens is much more advanced.

Although AR HMD has much potential, Google Glass still failed and left some unsolved question that so far HoloLens cannot give the solution, either.

As the pioneer of the software products around the world, Microsoft will fully support developing applications related to HoloLens, but as HoloLens is a brand new highly integrated product, the development of third party applications is a huge challenge. Because at first HoloLens is an independent product, so for the developer, it is a new platform. In addition, there are a lot of different sensors on HoloLens, so developers will not only think about the quality of the application, but also would want to make sure the interaction between human and machine. Finding the balance between virtualilty and reality, and creating an immersive experience are not easy to achieve. Secondly, refer to another product Kinect, there are also few developers who can make a high quality game.

Compared with Google Glass, HoloLens is quite large, as a wearable device, it should be worn outside. So there is a portability problem, Besides that, it is still unknown how the battery life is, whether there is a heating problem and whether it can satisfied shortsighted users. Last but not the least is the price of HoloLens and popularization of it. For example, in the case of Google Glass, the high prices and a high profile promotion are the main factors that it failed. At this stage people still define AR and VR as high-end electronic products and the market is for developers and fans.

There is not doubt that the research and development cost on HoloLens is far more than that of Google Glass. Thus it is hard to make it popular at least in the first iteration. What Microsoft should consider is how to make HoloLens more useful and on this basis try to persuade more consumers and potential users to buy this product.

5 MIXED REALITY

ADVANCES in augmented reality (AR) and virtual reality (VR) promise to change the way we interact with technology.

Though consumer buzz surrounding devices designed for gaming and entertainment is growing, AR and VR's enterprise potential is proving to be the real cause for excitement. Across sectors, use cases and concepts are emerging, and pilot programs are ramping into production.1

Meanwhile, the Internet of Things (IoT) is attracting more business investment as attention begins shifting from underlying sensors and connected devices to real-world scenarios driven by advances in IoT technology. Pioneering applications are emerging in the areas of personal health and wellness, supply chain, and in the civic infrastructure of smart cities, among others.2

Mixed reality (MR) represents the controlled collision of the AR/VR and IoT trends. With MR, the virtual and real worlds come together to create new environments in which both digital and physical objects—and their data—can coexist and interact with one another. MR shifts engagement patterns, allowing more natural and behavioral interfaces. These interfaces make it possible for users to immerse themselves in virtual worlds or "sandboxes," while at the

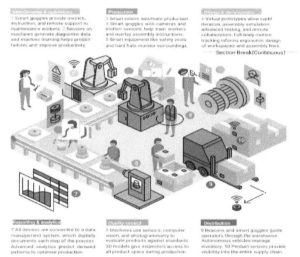

same time digesting and acting upon digital intelligence generated by sensors and connected assets. For example, as a worker wearing smart glasses examines a system in a remote location, diagnostic information appearing in his field of vision indicates the system is malfunctioning. If the worker can't fix the problem himself, skilled technicians in another location would be able to transmit detailed digital instructions for repairing the malfunction and, then, walk him through the repair process quickly and efficiently. In this and similar scenarios across industries and operating models, MR makes it possible to deliver actionable information to any location where work is done—on site, on the shop floor, or in the field.

The mixed-reality trend is being fueled by investments in platforms, devices, and software ecosystems. The ultimate goal of these investments is to replace keyboards and flat displays with entirely new paradigms for communication and collaboration. If successful, this would represent the biggest fundamental shift in user engagement we have in seen in

41

the modern technological era.

The where is the what

Mixed-reality use cases and patterns have emerged around early investments. Though specific priorities differ by industry, many land somewhere within the following areas:

Training, education, and learning: AR and VR can be used to onboard new employees and develop existing talent by immersing them in highly realistic, virtual work environments that feature both educational information and interactive problem-solving challenges. Immersive environments offer a number of advantages over traditional teaching methods. For example, they provide trainees with safe (virtual) exposure to complex and potentially dangerous equipment and scenarios. They also make it possible for supervisors to review video recordings of training sessions to monitor progress and tailor lessons to specific employee needs.

Operations: By providing field service technicians, warehouse pickers, assembly-line workers, and others with IoT applications and digital system content tailored to their unique tasks, companies may be able to boost employee productivity and streamline work processes. On the production floor, for example, job aids can guide workers performing kitting tasks to locations of shelved items. In the field, engineers could access the service history of specific equipment, guidance on triage and repair steps, and any real-time information that intelligent devices may be generating from a site. They would review this information in a hands-free, heads-up manner that maintains their autonomy and supports worker safety.3

Communication and collaboration: As organisations become "unbounded" by eliminating functional siloes, multidisciplinary teams will be able to work seamlessly together within and across company walls. Mixed reality can support this kind of next-generation interaction by replacing shared productivity tools and videoconferencing with immersion and a sense of presence. Workers in disparate locations can interact with the same digital artifacts, just as if they were in one conference room manipulating the same physical objects. Automotive manufacturers are applying these methods to visualise design improvements of existing components—dramatically accelerating the concept-to-manufacturing process among globally distributed teams.4 Likewise, research and development functions, construction and engineering firms, and even some services organisations are exploring this new style of collaboration, removing geo-temporal constraints from both creative processes and more tangible operations.5

Marketing and customer service: From high-tech experiential marketing to virtual branches, MR—specifically leveraging VR capabilities—can provide experiences that not only replicate aspects of the real world but build compelling engagement patterns predicated on the convergence of technologies. For example, you will be able to put a virtual product in your customer's hands and then guide her experience, gauge her responses, and personalise content based on this customer's transaction history and preferences.6

Shopping: Virtual reality stands poised to revolutionise the way we shop. Consider, for example, "walking" through a virtual cruise-ship cabin or hotel suite before booking it or immersing yourself in a virtual jewelry store where you try on necklaces that catch your eye. Mixed-reality capabilities could enhance these virtual experiences by providing pricing or descriptive information on the products or services you are considering, along with suggestions for similar offerings.7

MR technology: Enablers and obstacles

As it often does with emerging technologies, tremendous hype surrounds AR's and VR's promise—for good reason. These platforms offer new ways of framing and delivering content,
experiences, and interactions. They also bring with them opportunities to redefine the tools, models, and business processes that they could potentially replace. But as MR devices, software and standards evolve, enterprises are discovering hidden challenges beyond the management of technology.

AR, VR, and IoT represent new categories of devices that need to be managed and secured. Learning from the workplace adoption of smartphones and tablets, organisations can begin their MR journeys on the right foot by developing appropriate controls and policies to monitor and enforce enterprise needs. Security and privacy are important considerations—at the device level, the data and supporting content level, and the application level. Safety and regulatory compliance implications are paramount, especially since many potential scenarios involve critical infrastructure and sensitive operations.

To process event streams, render mixed-reality experiences, and capture and respond to the movements of an individual user, platforms will need several enabling environments. They must be immediately aware of that user's role, the context in which he works, and information he needs to complete his tasks. MR's ability to simultaneously track an environment and an individual's relationship to that environment is the key to allowing virtualised objects and information to respond realistically to what that individual is doing. This requires the kind of complex signal processing and response generation found in some advanced videogame designs. It's no surprise that early experiments have been built on top of popular gaming development engines such as Unreal.8

MR must also navigate interfaces that allow for the blend of voice, body, and object positioning to open up a world of business process opportunities in every industry. Recognising subtle expressions requires precise techniques, from eye tracking to interpreting a wink or a smile correctly. MR design patterns should make it possible for digital content to react intuitively to signals. New vocabularies for design patterns are required, along with solutions to manifest in both the AR/VR systems, as well as the surrounding connected devices and sensor-enabled environments.

MR should bring together relevant data to provide insightful interactions. That could mean integrating MR with the core in order to access information residing in ERP, CRM, HR, marketing, and other systems. Conversely, MR can be an important tool to digitise work—potentially automating underlying workload, updating data with supplemental supporting information, or allowing MR steps to be a part of a longer-running business process.

The context of content

Perhaps most importantly, natural interfaces make it possible to simultaneously process the meaning, intent, and implications of content in context of how the world works—and how we behave in it.

A digital object's weight, scale, angle, position, and relationships with other virtual and real-world objects allow designers to fashion truly visceral experiences. Sound and haptics (or sensory feedback) can replace graphical predecessors, allowing for the enterprise to rethink alerts, warnings, or the completion of tasks. Built-in object and facial recognition gives us the ability to map spaces and people to accurately render in the virtual or augmented experience, and to insert purely digital enhancements in a way that seems both natural and realistic.

MR requires an entirely new set of digital content and context. High-definition, 360-degree renderings of facilities and equipment are necessary to translate the real world into virtual environments or to marry augmented physical and digital experiences. Sensors and embedded beacons may also be required to track devices, equipment, goods, and people. Likewise, meta-data describes not just an asset's base specifications but also its composition, behavior, and usage—all necessary to simulate interactions.

All together now

Even as we elevate mixed reality above its piece-parts of AR, VR, and IoT, the underlying technologies themselves are continuing to advance. Individually they represent very different solutions, domains,and potential applications. However, if companies pursue them in isolation, their full potential will likely never be realised. The goal should be evolving engagement—building more intuitive, immersive, and empowering experiences that augment and amplify individual users, leading to new levels of customer intimacy, and creating new solutions to reshape how employees think and feel about work. If done correctly, mixed reality may open floodgates for transforming how tomorrow's enterprises are built and operated.

Using immersive technologies to protect child welfare

Each year, newly graduated social workers enter child welfare agencies determined to make a difference. Unfortunately, many of them may be unprepared, both practically and emotionally, for what awaits. It can take years of on-site visits to homes and childcare centres to help a social worker develop the deep observational skills and attention to detail required to accurately assess a child's living situation and, then, to determine whether further investigation and action is necessary.

Immersive technologies may someday offer child welfare agencies an efficient way to accelerate that learning process. Using serious games— games designed specifically to teach skills—and 3D simulation, social workers would be able to practice real-time engagements designed to help develop sensitivities and nuanced evaluation skills previously achieved only after years on the job.

For example, one training module might place a trainee social worker in a virtual home setting and ask her to identify all potential risk factors in a few minutes. After the initial scan, the trainee returns to the virtual room, where risk factors she overlooked during the initial assessment are blinking. When the trainee points at a specific signal, a description of the specific risk factor and an explanation of its importance appear in her field of vision. As part of this same process, the trainee could also practice reacting to difficult situations and documenting what she sees.

3D training models could be customised and refined for use in any environment or scenario. Not only could this expand the breadth of training available for new recruits—it would likely help veteran social workers further their professional development. When used over time, these capabilities might also help agencies assess and improve their overall effectiveness and teach critical thinking and decision making. In this light, MR capabilities are not just technical game changers but behavioral solutions, creating experiences that potentially benefit social workers and—even more importantly—help those they serve.

AR meets the IoT on the shop floor

In 2016, two innovative companies introduced to each other at the MIT Media Lab, convened at Jabil Blue Sky, an innovation centre in Silicon Valley to kick the tires on a new digital manufacturing process technology. The companies involved were Jabil, a global provider of engineering, manufacturing, and intelligent supply chain solutions, and Tulip, which offers a cloud-based platform featuring shop-floor apps, industrial IoT, and real-time analytics.
The solution being tested? A new cloud-based operating system that feeds IoT production-line data in real time to workers on a shop floor through their smartphones and tablets. By monitoring this information stream as they perform their production tasks, workers can respond on the fly to process changes. Eventually, the system could also integrate the power of mixed reality into manufacturing environments. The goal is to improve manufacturing flow through the shop floor. With real-time information enabled by the latest improvements in digital technology, companies may be able to reduce worker pauses or idle machine cycles that typically accompany changes in production conditions.

After initial testing at the Blue Sky innovation centre, Jabil and Tulip deployed the system in a production environment used by workers executing highly specialised work processes. For a period of six months, engineers monitored cycle and step-times data to further optimise manufacturing processes through continuous time studies and root-cause analysis exercise aided by the new digital tools. The results? Production yield increased by more than 10 percent, and manual assembly quality issues were reduced by 60 percent in the initial four weeks of operation, which exceeded customer accepted yields and predictions for the current design.

According to Tulip co-founder Natan Linder, in the near future, augmented reality (AR) capabilities will likely amplify the power of IoT manufacturing solutions such as the one his company tested with Jabil. The delivery of contextual information to workers without requiring a screen is already providing significant benefits to global manufacturers, says Linder, citing increased product and service quality, increased worker productivity through reduced rework, and higher throughput, as well as reduced training time. "Increasingly, we're seeing deployment of light-based AR in manufacturing, which uses lasers and projectors to layer visual information onto physical objects. This approach doesn't have many of the disadvantages of other AR interfaces; most importantly, it doesn't require workers to wear headsets.

"The real power of augmented reality comes into play," he continues, "when it is combined with sensors, machines, and data from smart tools. These IoT data sources provide the real-time information that the hands-on workforce needs to get work done and optimise processes, with augmented reality delivering the information at the right time and in the right place."

Yeah, but can you dance to it?

By adding production capabilities to data collected from sensors and multiple cameras in the field, mixed-reality solutions can transform how humans interact (visually and socially) with the world around them and the events they "attend."

In the last few years, we've seen VR broadcasts of sporting events as diverse as NASCAR, basketball, golf, and even surfing. Those initial broadcasts gave viewers a 360-degree view of the playing field and allowed them to choose their own vantage point throughout the game or race, supplemented with fully mixed 3D VR audio, announcer commentary, VR-like graphics, and real-time stats.

The use of sensors can enable broadcasters to provide additional value to remote viewers. Spanish start-up FirstV1sion, a wearables company that embeds video and radio transmission equipment in athletes' uniforms so VR viewers can watch a play or a game from a specific player's view point, outfitted players for European soccer matches and basketball games. In addition to the video feed, the electronics include a heart monitor and accelerometer so viewers can track players' biometric data as the game progresses. The hope is that fans will be more emotionally involved in the game if they can see a player's heart rate increase as the action heats up.

Musical events are obvious settings for social interaction, and virtual streaming of concerts is becoming more common. For example, the Coachella festival provided a cardboard headset with each ticket sold last year in case concertgoers didn't want to leave the hospitality tent to watch a band. One VR company is taking it a step further to let music fans be the performers: TheWaveVR has developed a virtual reality concert platform that not only allows users to watch musical performances— when paired with the HTC Vive, it enables them to DJ their own set in a virtual venue. Viewers can listen and dance to the music in the "club" while talking to other attendees.

On the political front, NBC News used mixed reality to encourage dialogue between American voters during the recent presidential campaign. It virtually recreated the real-life "Democracy Plaza" it had erected at New York's Rockefeller Centre so viewers located anywhere could enter the plaza, view live programming, access real-time viewer opinion polls, interact with its newscasters and pundits, and, most importantly, engage with other audience members.

It seems the dire predictions of virtual reality's isolationism have been much exaggerated. By adding sensors, voice recognition, and data overlays to create a mixed reality in which humans can interact more naturally, the future looks quite engaging for playtime as well as industry.

6 IMPLEMENTING AUGMENTED REALITY IN LEARNING

In today's world, technology has become a crucial part of our lives. It has changed how people think and apply knowledge. One of the newest developing technologies is augmented reality (AR), which can be applied to computers, tablets, and smartphones. AR affords the ability to overlay images, text, video, and audio components onto existing images or space. AR technology has gained a following in the educational market for its ability to bridge gaps and bring a more tangible approach to learning. Learner-centered activities are enhanced by the incorporation of virtual and real-world experience. AR has the potential to change education to become more efficient in the same way that computers and Internet have.

RESEARCH

Picture 23-a [1]

Research conducted for this literature review focused on learning applications of AR. The initial search of K-12 applications was far too broad to provide a valuable synthesis. The keywords included learning applications, science or STEM focus, and augmented reality. Journals with a concentration in technology and education that held significance to AR within the classroom setting were sought. References were included that explained the concept of AR as well as studies that implemented AR. Most of the references for this analysis were published within the past five years; however, a few articles included were published as early as 2010.

THEORETICAL FOUNDATIONS

AR educational programs are learner-centered and related to learner interests. It allows learners to explore the world in an interactive way. Constructivism also encourages learners to work collaboratively, and AR provides learners the opportunity to do this in a traditional school setting as well as in distance education. Dunleavy et al. (2009) believe that the engagement of the learner as well as their identity as a learner is formed by participating in collaborative groups and communities. Constructivism has also changed the role of the educator to become a facilitator, where the responsibility to organize, synthesize, and analyze content information is in the hands of the learner (DeLucia et al., 2012).

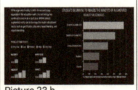

Picture 23 b

"Flow theory describes how people who are engaged in meaningful activities are more likely to stay focused. Bressler and Bodzin (2013) investigated a science gaming experience in relation to flow experience. Their study had a mean flow experience score of 82.4%, which indicates that the average learner experienced flow throughout the science mystery game that they played on an smart phone. This particular type of AR, as well as various others, connects their real-world surroundings to learning in a new and engaging way.

IMPLEMENTING AR IN LEARNING

AR allows flexibility in use that is attractive to education. AR technology can be utilized through a variety of mediums including desktops, mobile devices, and smartphones. The technology is portable and adaptable to a variety of scenarios. AR can be used to enhance content and instruction within the traditional classroom, supplement instruction in the special education classroom, extend content into the world outside the classroom, and be combined with other technologies to enrich their individual applications.

Traditional classroom uses

In any educational setting, there are often limitations in the various resources available. This is often seen foremost in the traditional classroom. Due to budget restraints or constraints on time, the means to teach learners in scenarios that allow them to learn by doing can be a challenge. Desktop AR allows learners to combine both real and computer-generated images. Iordache and Pribeanu (2009) used desktop AR that combined a screen, glasses, headphones, and a pointing device that allowed learners to conduct a hands-on exploration of a real object, in this case a flat torso, with superimposed virtual images. Computer images could show the process, but the pointing device allowed learners to guide their learning.

Classrooms can shift from the traditional lecture style setting to one that is more lab and learner-oriented. A case study conducted with a visual arts class noted that allowing learners to freely explore a room that was set up with webcams and desktops encouraged more activity while the learners perceived that they were more motivated to learn (Serio et al., 2013).

Instead of receiving information via images and lecture, learners had access to multimodal representations including text, audio, video, and 3D models.

Special Learning Uses

Because of the variety of tools that can be overlaid in an augmented environment, learners with physical disabilities can benefit from the potential learning aides that could be incorporated. Something as simple as overlaying audio for those with visual impairments or text for those with hearing disabilities can be effective tools when considering disability access (Forsyth, 2011). Head-mounted displays (HMD) can provide a hands-free device to project the overlay visuals to a learner and adjust the images based on the orientation of the learner while other devices enable learners to interact with the environment via voice recognition, gesture recognition, gaze tracking, and speech recognition. Bringing this technology to the classroom has the potential to allow for differentiated instruction and enrichment of the learning experience of learners with special needs. Evaluation trials conducted by Arvantis et al. (2009) showed that using wearable AR technology with learners who had physical disabilities produced, "interestingly comparable results with able-bodied users," (p. 250) in terms of "wearability" and pedagogy.

Outside the Classroom

Mobile applications can extend the traditional classroom beyond the physical walls. Annetta, Burton, Frazier, Cheng, and Chmiel (2018) reported that the percentage of 12 to 17 year olds who have their own mobile device is 75%, compared to 45% in 2004, and regardless of a learner's socioeconomic status, the number of learners carrying their own mobile devices is growing exponentially every year. Camera phones and smartphones allow users to gather information in a variety of locations. QR codes and GPS coordinates can be used to track and guide movement of the learners. Although several researchers chose to take learners off campus and conduct investigations in a field trip setting, others chose to remain within the grounds of the school.

An important point to note from this research is that GPS will not work inside of buildings. Therefore, any indoor activity would need to be conducted without a location-based AR technology.

CHALLENGES

Training

Training is an important aspect of AR. "Most educational AR systems are single-use prototypes for specific projects, so it is difficult to generalize evaluation results" (Billinghurst & Dunser, 2012, p. 61). Educators did not feel confident when setting up or implementing the program. In addition, educators who are normally lecture focused had a hard time letting go and allowing learners to explore the learning environment on their own.

A training should be provided for educators to learn a hands-off approach with their learners and show them how this way of teaching will foster an effective learning environment. The fear of not knowing what is on each learner's device can be elevated according to the authors through the process of allowing the learners more control over their learning.

Technical Problems

Dunleavy et al. (2009) showed that the GPS failed 15-30% during the study. A GPS error refers to either the software of the GPS itself or incorrect setup. This was considered the "most significant" malfunction. Other malfunctions identified in this study were the ability for the devices to be effectively used outdoors. The glare from the sun as well as the noisy environment could impair the learning of the learners.

There are several different kinds of devices that can be used when implementing AR in the classroom. Glasses, hand-held devices, and headwear are ways for the user to see computer -generated images imprinted on their reality.

Acquiring devices that are calibration free or auto-calibrated can be beneficial to the user as to avoid malfunction and user frustration.

Learner Issues

One issue identified in Dunleavy et al. (2009) determined that some AR situations can be dangerous. In this particular Alien Contact! scenario, learners must look at their handheld devices to participate. When engaging in activities outdoors the learners are unable to work on their devices and watch where they are going simultaneously. Therefore, learners were found to be wandering into roadways and needed to be redirected to safety by educators.

REACTIONS

Learners

Overall, learners reacted positively to using AR technology both in and outside of the classroom.

Learners also reported that learning in an AR environment is more stimulating and appealing than viewing a traditional slide presentation (i.e., Microsoft PowerPoint, SmartNotebook) because they preferred the audio, video, and feeling as if they were part of the 3D model that was transposed into a real physical space (Serio et al., 2013).

Educators

Educators may feel alarmed as if AR will "overtake" their classrooms; it seems that once learners experience this type of learning, they will not go back to their previous ways of learning. However, Annetta et al. (2012) expresseed that AR can be an activity to engage learners in future units and discussions. Billinghurst and Dunser (2012) believe that AR is a new form of face-to-face instruction, as learners share the learning experience. Educators have reported learners taking responsibility

and ownership of their learning (Kamarainen et al., 2013). Therefore, educators using AR technology are becoming facilitators to their learners.

IMPLICATIONS FOR RESEARCH

The importance of this literature review is that it not only showcases the current trends in AR technology but also its focus on the increased research and potential further application in the educational setting. Several components remain to be explored. When using AR outside of the classroom, educators and learners are able to use this as a tool for physical activity (Dunleavy et al., 2009). Linking learning with exercise and activity in an educational way can improve the perception that technology creates a non-interactive environment (NAEYC & Fred Rogers Center, 2012).

Another is that the amount of visual information that can be displayed on the screen can be overwhelming to learners. Studies should further explore the effects AR has on cognitive load in the brain and how much information should be displayed before it turns from a beneficial device into a distracting device (Bressler & Bodzin, 2013; Van Krevelen & Poelman, 2010).

7 CREATING AUGMENTED REALITY

To augment reality is to alter the view of the physical world through use of computer-generated sensory and image processing. It is the combination of reality and virtual reality, and acts as a technological extension to our own vision. The uses of augmented reality range from displaying critical information about a patient during surgery, to showing which the highest rated restaurant in a food court is. Information being overlaid on top of reality in real-time could drastically improve the efficiency and effectiveness of almost any every-day activity. This technology is still at its breaking through point and has much progress to make. This document will only skim the surface of augmented reality technology, but still make an effective use of the technology on a small rover.

D'Fusion

Source: Total Immersion website

1. The user presents an image to a webcam that is used as a connector to the real world

2. The image is recognized in the real-time video flow captured from the webcam

3. A 3D computer-generated object is then superimposed on the image as seen in the captured video

4. The user can then interact with the 3D object, in real-time, by moving the image in the real world

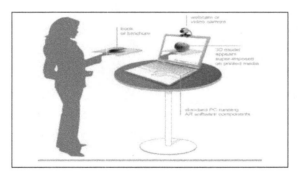

Augmented Reality Engine:

- A calibrating tool (D'Fusion Camera Calibration) to calibrate camera and sensors

- An authoring tool embedded in Autodesk Maya and 3ds Max (D'Fusion Exporter for Maya, D'Fusion Exporter for 3ds Max and D'Fusion 3D Viewer) to produce 3D content, preview it, and export it for D'Fusion's real-time engine

- An authoring tool (D'Fusion Studio) to design your scenarios, and define behaviors and interactions that can be script-controlled using the Lua scripting language

- A scenario engine (D'Fusion @Home, D'Fusion Mobile) to manage the real-time show

- A physics engine (based on the Bullet engine) to improve natural interactions and

rendering

- A rendering engine (based on the Ogre3D engine) to smoothly merge computer-generated objects with the real world through a real-time video stream

- A debugger to ensure a smooth development process

- An open framework, to easily add your own plug-ins and match your specific needs

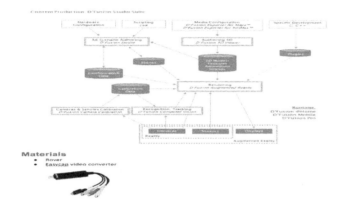

Materials
- Rover
- Easycap video converter

- Rover

- Easycap video converter

- Spycamera

- Blender and D'fusion software

Procedure

The first step to giving the rover an augmented reality is to first give it a reality. This involves installing the camera both physically on the rover, and intangibly using drivers for the camera. Since the USB driver for the receiver is not a "video capture device", a video converter must be used. This converter will take the analog video from the camera receiver, and convert it to a usable digital format through USB. If using windows 7, the automatic driver retrieval will not work for the Easycap converter. This driver will work to access the cameras.

Next we must install blender and the necessary resources to export the proper file. It is important to note that Blender version 2.62 must be installed, not the newest one. There is a newer beta version of the specialty exporter that should work with the newer versions of blender, but the trusted older version used in this demonstration only works with Blender 2.62. After installing blender, the Ogre exporter needs to be installed. Both the old version and the beta versions of the exporter can be found here. There are two methods of installing these add-ons. The first method is to use blenders interface. Under user-prefs, click add-ons, click 'install-add-on', and select io_export_ogreDotScene.py. The second method is to copy io_export_ogreDotScene.py to your blender installation under blender/2.59/scripts/addons_contrib/. Once the add-on has successfully been implemented, there should be a new option under file/export/ called Ogre3d(.scene and .mesh). This add-on will export the

3d model as the correct .mesh file needed for D'fusion to operate. For this exporter to work, it requires

OgreCommandLineTools. This should be installed to the C:/ directory. You may also need to install OgreMeshy to the C:/ directory for the exporter to work. The next step involves making your 3d rock for the rover.

Creating a 3D object using Blender

Blender is a very useful program for creating 3D objects to use for our augmented reality program. It is probably useful to watch the "getting started" videos on the cg cookie website: Get started with Blender
For our rock, we delete the default box mesh, and add a new mesh (icosphere).

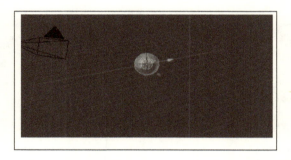

We then add a modifier called displacement to the rock. This will displace the vertices of the icosphere which will give us a random shape to work with.

After this you can smooth the edges by using the a subdivision modifier with 5-6 subdivisions on the object. Note: if you plan on adding a texture to create your rock, you should add the modifier, but do not click apply. This will cause issues when unwrapping the object for texturing.

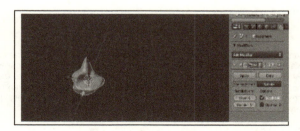

We then use the texturing method used in this youtube video: Rock texture tutorial. The only difference is we take the rock as a reddish/orangish color to make it resemble a rock from Mars.

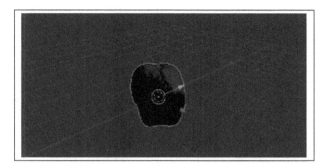

Note: texturing is affected by the lighting on the object. this can be avoided if you use normal mapping (instead of using full render that was used on this object)like explained in the above tutorial video.

We can also use the sculpting method using Alphas and different sculpting brushes. This method is much more difficult and is harder to make the rock look more realistic. However, you don't have to bother with loading the texture file into the D'fusion program. The object itself is modified. Note: For this method, you must click apply on the subdivison modifier so that the planets of the object will be subdivided very small so that you can sculpt each plane of the object.

Other useful videos for learning how to create your own rocks and texturing:

Useful tutorials on creating different objects:CG Cookie

Multi-textures and sculpting: Damaged pillar tutorial

Creating your Augmented Reality

After exporting the 3d object using the Ogre exporter, This portion takes your 3d object created in blender, and digitally places it on top of a tangible picture of your choosing. The D'fusion website offers a very in depth and easy to understand tutorial, so we will use that as a reference and explain what to change or do differently. Here is the tutorial

Step 1 and 2

The project name could be whatever you desire, but the options selected should all be the same as the tutorial.

Step 3 and 4

Select the USB2.0ATV driver, which is the receiver for the rover's onboard camera.

Step 5

Instead of importing their DemoRobot folder, select the folder that the 3d object was exported to. Drag the .mesh file into the 3d view instead of their DemoRobot.scene file. The positioning should be the same as their demo, so use their numbers. This is our rock without the texture, more on how to texture it later.

Step 6

There is no need to open the animation window, since the rock will not be animated. The

"robot_manager" files created could be renamed as "rock_manager", and select "entity" as the owner.

The rock_manager.lua script will not have any animation, so instead of their 4 lines of code ours will be:

```
local rock = scenette(getCurrentScriptOwner())
```

Step 7

The tutorial .zip file they ask you to download contains the "tracking.lua" file needed to make the image tracking function correctly. If not done so already, this file should be downloaded and added to the project.

Step 8

When creating the .bat file make sure to change the name "my_scenario.dpd" to the name of your project, also keep the quotation marks around the entire line.

Texture

Since the blender exporting does not include the texture, we need to add it through D'Fusion. To do this, we must creaete a new lua script and wrap a JPEG around the 3d object. Import the JPEG as by going to add>2d elements>texture. Name this new texture as texture1. Next add a lua script to the scene by going to add>script. Put this code inside the script file.

```
local scene = getCurrentScene()

local me = Entity(scene:getObjectByName("entity")) local mat0 = Material(me:getMaterial(0))

local tex1 = Texture(scene:getObjectByName("texture1")) mat0:setTexture(tex1)
```

This should wrap your object in the texture of your choosing.

Augmented reality can help in everyday life with many aspects. With AR we are able to simulate more accurately what a rover mission on mars would look like. The plaques for our mission can be replaced completely with 3D objects that look like mars rocks.Some issues that we encountered with our project, is that the camera quality was not perfect so if the plaque was too far away it wouldn't be able to pick up the image on the plaque clearly enough to create the 3D object on top of it. We also encountered a few problems with the exporting of the 3D object. This was easily fixed by changing our 3D modeling program from version 2.63 to 2.62. We can improve this project by creating multiple rocks for our project and multiple images to put the rocks onto (We found a few links that explain using multiple 3D objects instead of just one object). There is also a way to create a background landscape that we can fill with red sand/dirt.

8 CONCLUSION

In conclusion, as a bold try of the next stage of human-computer interaction, virtual reality and augmented reality now draw people's attention. Several high tech companies start to develop these kinds of devices one after another. From the analysis, we can see that the future of HCI is to let people more free, bring back a natural way to gain information, let people rely on the simpler things. At this point, many people may say usability, but in fact it is not good, because usability is from the angle of the product. What is more important is to put people first.

There are two ways to achieve the freedom that people expect. The first, one is the virtual way, people put their mind into a vast virtual space and release it. A pure digital spirit can be built, people get their freedom and this is what virtual reality technology wants to do. This is easy to achieve, so many companies develop virtual reality and there are some great products on the market. However, this direction is not that ideal, no matter how real the virtual environment is. After all it is a lie and people will find a way to go back to the real world. Thus virtual reality is more popular in games and design industries. In the long term, augmented reality will become a ubiquitous technology, It breaks the limitation of space and time, combines the digital content and real world, people's body and spirit can get fully used. Technology strengthens our body and serves our mind, people can live in a free and real world, hence the augmented reality can be used in many different industries.

Both virtual reality and augmented reality have years of history. Virtual reality even has more than 40 years history. There are two main reasons why they are not popular at present. Firstly, from a scientific and technological point of view, there are many limitations that need to be solved, like the battery life, power of the processor, motion-sickness problem with the HMD. The other reason is the rigid demand from the people whether they really need this technology in their daily life, or it is only needed in some the fields.

All in all, as a revolutionary technology, although there are some barriers between virtual reality and augmented reality, Microsoft, Google, Oculus Rift and other companies, those great development teams, never give up finding the next way of HCI. Even if those advanced products disappear in the history, they still have the unforgettable value, to lead the industry step into a new time. From the PC to the smartphone, we saw the new interaction come, from smartphone to the wearable device, we will look forward to the next way of interaction.

REFERENCES

Abowd, G. D., and Beale, R. (1991). 'Users systems and interfaces: a unifying framework for

interaction. In: D. Diaper and N. Hammond. Eds: HCI'91: People and Computers VI, pages 73-87. Cambridge: Cambridge University Press.

Abrash, M. (2012). [Blog] Available at: http://blogs.valvesoftware.com/abrash/latency-the-sine-qua-non-of-ar-and-vr/

Azuma, R A Survey of Augmented Reality Presence: Teleoperators and Virtual Environments, pp. 355–385, August 1997.

Broadband Backhaul. (2014). 1st ed. [ebook] Ceragon Networks Ltd. Available at: http://www.ceragon.com/images/Reasource_Center/Technical_Briefs/Ceragon_Technical_Bri ef_Asymmetric_Transport.pdf.

Docs.opencv.org, (2014). Camera Calibration and 3D Reconstruction — OpenCV 2.4.11.0 documentation. [online] Available at http://docs.opencv.org/modules/calib3d/doc/camera_calibration_and_3d_reconstruction.ht ml [Accessed 7 Apr. 2015].

Edwards, B. (2008). The computer mouse turns 40. [online] Macworld. Available at: http://www.macworld.com/article/1137400/mouse40.html

Google Inc., (2013). Lightweight eyepiece for head mounted display. US20130070338 A1.

Google Inc., (2013). Heads-up display including eye tracking. US 20130207887 A1.

Goldman, David (April 4, 2012). "Google unveils 'Project Glass' virtual-reality glasses". Money (CNN).

Hewett; Baecker; Card; Carey; Gasen; Mantei; Perlman; Strong; Verplank. (2009) "ACM SIGCHI Curricula for Human-Computer Interaction". ACM SIGCHI.

Heim, M. (1993). The metaphysics of virtual reality. New York: Oxford University Press.

Hempel, J. (2015). Project HoloLens: Our Exclusive Hands-On With Microsoft's Holographic Goggles | WIRED. [online] WIRED. Available at: http://www.wired.com/2015/01/microsoft-hands-on/.

iFixit, (2014). Oculus Rift Development Kit 2 Teardown. [online] Available at: https://www.ifixit.com/Teardown/Oculus+Rift+Development+Kit+2+Teardown/2763

Izadi, S., Kim, D., Hilliges, O., Molyneaux, D., Newcombe, R., Kohli, P., ... & Fitzgibbon, A. (2011). KinectFusion: real -time 3D reconstruction and interaction using a moving depth camera. In Proceedings of the 24th annual ACM symposium on User interface software and technology (pp. 559-568). ACM.

Johnson, E.A. (1965). "Touch Display - A novel input/output device for computers". Electronics Letters 1 (8): 219–220.

Kickstarter, (2012). Oculus Rift: Step Into the Game. [online] Available at: https://www.kickstarter.com/projects/1523379957/oculus-rift-step-into-the-game.

Klein, G. and Murray, D. (2007). 1st ed. [ebook] Available at: http://www.robots.ox.ac.uk/~gk/publications/KleinMurray2007ISMAR.pdf .

Lee, Kangdon (March 2012). "Augmented Reality in Education and Training". Techtrends:

Linking Research & Practice To Improve Learning 56 (2). Retrieved 2014-05-15.

Licklider, J. (1960). Man-Computer Symbiosis. IRE Transactions on Human Factors in Electronics, HFE-1(1), pp.4-11.

Norman, D. (1988). The psychology of everyday things. New York: Basic Books.

Oculus VR, (2014). The All New Oculus Rift Development Kit 2 (DK2) Virtual Reality Headset. [online] Available at: https://www.oculus.com/dk2/.

OCULUS VR, LLC, (2014).1st ed. [e-book] OCULUS VR, LLC. Available at: http://static.oculus.com/sdk-downloads/documents/Oculus_Developer_Guide_0.4.4.pdf.

P. Milgram and A. F. Kishino, Taxonomy of Mixed Reality Visual Displays IEICE Transactions on Information and Systems, E77-D(12), pp. 1321-1329, 1994

Riftenabled.com, (2015). RiftEnabled™ – Oculus Rift enabled list of games & demos. [online] Available at: http://www.riftenabled.com/admin/apps/.

Shackel, B. (1959) Ergonomics for a computer. Design 120, 36-39.

Sutherland, I. E. (1968). "A head-mounted three dimensional display". Proceedings of AFIPS 68, pp. 757-764

Summers, N. (2013). 10 Experimental Projects that Completely Rethink Computer Interfaces. [online] The Next Web. Available at: http://thenextweb.com/insider/2013/05/28/instinctive-innovation-10-experimental-projects-that-completely-rethink-computer-interfaces/

Yin-Poole, W. (2015). Watch Valve's HTC Vive VR Rezzed developer session. [online] Eurogamer.net. Available at: http://www.eurogamer.net/articles/2015-03-13-watch-valves-htc-vive-vr-rezzed-developer-session.

Annetta, L., Burton, E. P., Frazier, W., Cheng, R., & Chmiel, M. (2012). Augmented reality games:

Using technology on a budget. Science Scope, 36(3), 54-60.

Arvanitis, T. N., Petrou, A., Knight, J. F., Savas, S., Sotiriou, S., Gargalakos, M., & Gialouri, E. (2009).

Human factors and qualitative pedagogical evaluation of a mobile augmented reality system for science education used by learners with physical disabilities. Personal and Ubiquitous Computing, 13(3), 243-250.

Azuma, R., Baillot, Y., Behringer, R., Feiner, S., Julier, S., & MacIntyre, B. (2001). Recent advances in augmented reality. Computer Graphics and Applications, IEEE , 21(6), 34-47.

Benford, S., Anastasi, R., Flintham, M., Greenhalgh, C., Tan- davanitj, N., Adams, M., & Row-Farr, J. (2003). Coping with uncertainty in a location-based game. IEEE Pervasive Computing, 2(3), 34–41.

Billinghurst, M., & Dunser, A. (2012). Augmented reality in the classroom. Computer, 45(7), 56-63.

Bressler, D. M., & Bodzin, A. M. (2013). A mixed methods assessment of learners' flow experience during a mobile augmented reality science game. Journal of Computer Assisted Learning, 29(6), 505-517. doi: 10.1111/jal.12008

Collins, A., & Halverston, R. (2009). Rethinking education in the age of technology: The digital revolution and schooling in America. New York: Educators College Press.

DeLucia, A., Francese, R., Passero, I., & Tortoza, G. (2012). A collaborative augmented campus based on location-aware mobile technology. International Journal of Distance Education Technologies, 10(1), 55-71.
http://dx.doi.org.ezproxy.liberty.edu:2048/10.4018/jdet.2012010104

DNews. (2013, February 20). Google glass and augmented reality's future. Retrieved from http:// youtu.be/qdD5-woi_os

Dunleavy, M., Dede, C., & Mitchell, R. (2009). Affordances and limitations of immersive participatory augmented reality simulations for teaching and learning. Journal of Science Education and Technology, 18(1), 7-22.

Enyedy, N., Danish, J. A., Delacruz, G., & Kumar, M. (2012). Learning physics through play in an augmented reality environment. International Journal of Computer-Supported Collaborative Learning, 7(3), 347-378. doi:http://dx.doi.org/10.1007/s11412-012-9150-3

Forsyth, E. (2011). Ar u feeling appy? augmented reality, apps and mobile access to local studies information. Australasian Public Libraries and Information Services, 24(3), 125.

Goodrich, R. (2013, May 29). What is augmented reality? Retrieved from Goodrich, R. (2013). What is augmented reality? Retrieved from http://www.livescience.com/34843-augmented-reality.html

Iordache, D. D., & Pribeanu, C. (2009). A comparison of quantitative and qualitative data from a formative usability evaluation of an augmented reality learning scenario. Informatica Economica, 13(3), 67-74.

Kamarainen, A. M., Metcalf, S., Grotzer, T., Browne, A., Mazzuca, D., Tutwiler, M.S., & Dede, C. (2013). EcoMOBILE: Integrating augmented reality and probeware with environmental education field trips. Computers & Education, 68, 545-556.
doi:10.1016/j.compedu.2013.02.018

Morrison, A., Mulloni, A., Lemmela, S., Oulasvirta, A., Jacucci, G., Peltonen, P., Schmalstieg, D., & Regenbrecht, H. (2011). Collaborative use of mobile augmented reality with paper maps. Computers & Graphics, 35(4), 789-799.

Mullen, T. (2011). Prototyping augmented reality. Indianapolis, IN: John Wiley & Sons, Inc. NAEYC, & Fred Rogers Center, (2012). Technology and interactive media as tools in early childhood programs serving children from birth through age 8. Retrieved from http://www.naeyc.org/files/ naeyc/file/positions/PS_technology_WEB2.pdf

Rigby, C. S., & Przybylski, A. K. (2009). Virtual worlds and the learner hero: How today's video games can inform tomorrow's digital learning environments. Theory and Research in Education, 7(2), 214-223.

Serio, A. D., Ibanez, M. B., & Carlos, D. K. (2013). Impact of an augmented reality system on learners' motivation for a visual art course. Computers & Education, 68, 586-596.
http://dx.doi. org/10.1016/j.compedu.2012.03.002

Van Krevelen, D. W. F., & Poelman, R. (2010). A survey of augmented reality technologies, applications and limitations. The International Journal of Virtual Reality, 9(2), 1-20. Retrieved from
http://kjcomps.6te.net/upload/paper1%20.pdf

Wang, X. (2012). Augmented reality: A new way of augmented learning. eLearn, 10.
doi:10.1145/2380716.2380717

Wither, J., Tsai, Y., & Azuma, R. (2011). Indirect augmented reality. Computers & Graphics, 35(4), 810-822.

NELSON KUNKEL - Deloitte Digital US creative director, Deloitte Consulting LLP, https://www2.deloitte.com/tl/en/pages/technology/articles/tech-trends.html

STEVE SOECHTIG - Deloitte Digital Experience practice leader, Deloitte Consulting LLP, https://www2.deloitte.com/tl/en/pages/technology/articles/tech-trends.html